QU 68
T47
1997
c.4

KATHLEEN WILLIAMS

Therapeutic Protein and Peptide Formulation and Delivery

ACS SYMPOSIUM SERIES **675**

Therapeutic Protein and Peptide Formulation and Delivery

Zahra Shahrokh, EDITOR
Genentech, Inc.

Victoria Sluzky, EDITOR
COR Therapeutics, Inc.

Jeffrey L. Cleland, EDITOR
Genentech, Inc.

Steven J. Shire, EDITOR
Genentech, Inc.

Theodore W. Randolph, EDITOR
University of Colorado

Developed from a symposium sponsored by the
Division of Biochemical Technology

American Chemical Society, Washington, DC

Library of Congress Cataloging-in-Publication Data

Therapeutic protein and peptide formulation and delivery / Zahra Shahrokh, editor . . . [et al.].

    p.   cm.—(ACS symposium series, ISSN 0097–6156; 675)

"Developed from a symposium sponsored by the Division of Biochemical Technology at the 211th National Meeting of the American Chemical Society, New Orleans, Louisiana, March 24–28, 1996."

Includes bibliographical references and indexes.

ISBN 0–8412–3528–7

1. Protein drugs—Dosage forms—Congresses. 2. Peptide drugs—Dosage forms—Congresses.

I. Shahrokh, Zahra, 1958– . II. American Chemical Society. Division of Biochemical Technology. III. American Chemical Society. Meeting (211th : 1996: New Orleans, La.) IV. Series.

RS431.P75T47   1997
615′.19—dc21   97–22244
                                                                                         CIP

This book is printed on acid-free, recycled paper.

Copyright © 1997 American Chemical Society

All Rights Reserved. Reprographic copying beyond that permitted by Sections 107 or 108 of the U.S. Copyright Act is allowed for internal use only, provided that a per-chapter fee of $17.00 plus $0.25 per page is paid to the Copyright Clearance Center, Inc., 222 Rosewood Drive, Danvers, MA 01923, USA. Republication or reproduction for sale of pages in this book is permitted only under license from ACS. Direct these and other permissions requests to ACS Copyright Office, Publications Division, 1155 16th Street, N.W., Washington, DC 20036.

The citation of trade names and/or names of manufacturers in this publication is not to be construed as an endorsement or as approval by ACS of the commercial products or services referenced herein; nor should the mere reference herein to any drawing, specification, chemical process, or other data be regarded as a license or as a conveyance of any right or permission to the holder, reader, or any other person or corporation, to manufacture, reproduce, use, or sell any patented invention or copyrighted work that may in any way be related thereto. Registered names, trademarks, etc., used in this publication, even without specific indication thereof, are not to be considered unprotected by law.

PRINTED IN THE UNITED STATES OF AMERICA

Advisory Board

ACS Symposium Series

Mary E. Castellion
ChemEdit Company

Arthur B. Ellis
University of Wisconsin at Madison

Jeffrey S. Gaffney
Argonne National Laboratory

Gunda I. Georg
University of Kansas

Lawrence P. Klemann
Nabisco Foods Group

Richard N. Loeppky
University of Missouri

Cynthia A. Maryanoff
R. W. Johnson Pharmaceutical
    Research Institute

Roger A. Minear
University of Illinois
    at Urbana–Champaign

Omkaram Nalamasu
AT&T Bell Laboratories

Kinam Park
Purdue University

Katherine R. Porter
Duke University

Douglas A. Smith
The DAS Group, Inc.

Martin R. Tant
Eastman Chemical Co.

Michael D. Taylor
Parke-Davis Pharmaceutical
    Research

Leroy B. Townsend
University of Michigan

William C. Walker
DuPont Company

Foreword

THE ACS SYMPOSIUM SERIES was first published in 1974 to provide a mechanism for publishing symposia quickly in book form. The purpose of this series is to publish comprehensive books developed from symposia, which are usually "snapshots in time" of the current research being done on a topic, plus some review material on the topic. For this reason, it is necessary that the papers be published as quickly as possible.

Before a symposium-based book is put under contract, the proposed table of contents is reviewed for appropriateness to the topic and for comprehensiveness of the collection. Some papers are excluded at this point, and others are added to round out the scope of the volume. In addition, a draft of each paper is peer-reviewed prior to final acceptance or rejection. This anonymous review process is supervised by the organizer(s) of the symposium, who become the editor(s) of the book. The authors then revise their papers according to the recommendations of both the reviewers and the editors, prepare camera-ready copy, and submit the final papers to the editors, who check that all necessary revisions have been made.

As a rule, only original research papers and original review papers are included in the volumes. Verbatim reproductions of previously published papers are not accepted.

ACS BOOKS DEPARTMENT

Contents

Preface ix

1. Developing Pharmaceutical Protein Formulations: Assumptions
   and Analytical Tools 1
   Zahra Shahrokh

2. The Pharmaceutical Development of Insulin: Historical Perspectives
   and Future Directions 29
   Henry R. Costantino, Stanley Liauw, Samir Mitragotri, Robert Langer,
   Alexander M. Klibanov, and Victoria Sluzky

3. Stability of the Dipeptide Aspartame in Solids and Solutions 67
   Leonard N. Bell

4. Mechanisms of Methionine Oxidation in Peptides 79
   Christian Schöneich, Fang Zhao, Jian Yang, and Brian L. Miller

5. A Discussion of Limitations on the Use of Polymers for Stabilization
   of Proteins During the Freezing Portion of Lyophilization 90
   David M. Barbieri, Martin C. Heller, Theodore W. Randolph,
   and John F. Carpenter

6. Phase Separation and Crystallization of Components in Frozen Solutions:
   Effect of Molecular Compatibility Between Solutes 109
   Ken-ichi Izutsu, Sumie Yoshioka, and Shigeo Kojima

7. In Situ Formation of Polymer Matrices for Localized Drug Delivery 119
   Jennifer L. West

8. Transdermal Delivery of Macromolecules: Recent Advances
   by Modification of Skin's Barrier Properties 124
   Mark R. Prausnitz

9. Amino Acid Derived Polymers for Use in Controlled Delivery Systems
   of Peptides 154
   S. Brocchini, D. M. Schachter, and J. Kohn

10. Analysis of the Solution Behavior of Protein Pharmaceuticals by Laser Light Scattering Photometry 168
    Gay-May Wu, David Hummel, and Alan Herman

11. Applications of Ultraviolet Absorption Spectroscopy to the Analysis of Biopharmaceuticals 186
    Henryk Mach, Gautam Sanyal, David B. Volkin, and C. Russell Middaugh

12. Surfactant-Stabilized Protein Formulations: A Review of Protein–Surfactant Interactions and Novel Analytical Methodologies 206
    LaToya S. Jones, Narendra B. Bam, and Theodore W. Randolph

Author Index 223

Affiliation Index 223

Subject Index 223

Preface

EFFICIENT DEVELOPMENT OF NOVEL PROTEINS AND PEPTIDES as therapeutic agents requires a sound understanding of their physico-chemical properties, which are derived through the use of information-rich and sensitive analytical tools. Patient-convenient delivery of therapeutic peptides also continues to be a major challenge, and so a vast number of studies on novel, polymer-based, sustained delivery systems and targeted delivery systems have been initiated. Such technologies and issues are growing so rapidly that updated reviews on the subjects are continually needed.

This book is a compendium of some of the topics presented in three symposia at the 211th National Meeting of the American Chemical Society, sponsored by the ACS Division of Biochemical Technology, in New Orleans, Louisiana, March 24–28, 1996. Topics covered included protein and peptide formulation, delivery, and analytical advances. The book begins with a review of protein stability and analytical technologies. An example of nearly all the major issues of stability and delivery is then reviewed in the intensely studied case of insulin. The remaining chapters are divided into sections on specific protein and peptide degradation pathways, solid-state stability considerations, advances in delivery approaches, and analytical technologies.

Acknowledgments

We acknowledge the Division of Biochemical Technology and the Biotechnology Secretariat for sponsoring the symposia, particularly Govind Rao for his support. We thank Milianne Chin for invaluable administrative assistance in organizing manuscript distribution for review and submission to ACS Books. We also acknowledge Michelle Althuis, Vanessa Johnson-Evans, and Cheryl Shanks from ACS Books for their guidance in the publication process. The thorough and valuable inputs from the many reviewers are greatly appreciated.

ZAHRA SHAHROKH
JEFFREY L. CLELAND
STEVEN J. SHIRE
Genentech, Inc.
460 Point San Bruno Boulevard
South San Francisco, CA 94080

October 30, 1996

VICTORIA SLUZKY
COR Therapeutics, Inc.
256 East Grand Avenue
South San Francisco, CA 94080

THEODORE W. RANDOLPH
Department of Chemical Engineering
University of Colorado
Boulder, CO 80309–0424

Chapter 1

Developing Pharmaceutical Protein Formulations: Assumptions and Analytical Tools

Zahra Shahrokh

Department of Pharmaceutical Research and Development, Genentech, Inc., 460 Point San Bruno Boulevard, Mail Stop #82, South San Francisco, CA 94080

A marketable protein formulation must be safe to administer, remain physically, chemically, and biologically stable during the recommended shelf life, induce minimal local irritation, and meet the specific clinical and delivery needs. Achieving this goal is an evolutionary process involving many disciplines in process sciences, pharmacokinetics, toxicology, and clinical studies. With this scope in mind, this review will focus on protein stability and analytical tools for characterization which are critical to formulation development, demonstration of process robustness, and establishment of rational specifications. Of particular importance are the recent changes in the FDA requirements [1] that have allowed for a shift from showing in vivo biological equivalence in humans to demonstrating physical-chemical, bioactivity, and pharmacokinetic equivalence in animal models; this calls for greater understanding of the physical-biochemical stability of the protein using sensitive and specific analytical tools. In addition to providing a summary of the common protein degradation pathways in this review, theoretical considerations about protein folding and stability are outlined. Also, some underlying assumptions in early formulation selection are brought forth, and suggestions for efficient and rational formulation development are made.

© 1997 American Chemical Society

Temperature Dependence of Protein Stability - Theoretical Considerations:

Temperature has a significant effect on the conformational stability of a protein, as well as its chemical stability as with any chemical reaction. Conformational stability is a marginal difference of large contributions from electrostatics, van der Waals (or dipole interactions), hydrogen-bonding, and hydrophobic forces [2-5]. The effect of temperature on protein conformation is complex and controversial. Briefly, as temperature increases, there is a decrease in van der Waals, hydrogen bonding, and dipolar interactions [6, 7], with little change in electrostatic effects [3]. Hydrophobic effect, driven entropically near ambient temperatures, by ordering of water molecules around hydrophobic residues to minimize hydrogen bond breakage, increases with an increase in temperature [5]. It is believed that these hydrophobic forces drive thermal denaturation as a result of increased conformational entropy.

Calorimetric studies indicate endothermic thermal protein unfolding with a mid-point of denaturation, T_m, which can be as low as 35-40 °C (e.g. in acidic fibroblast growth factor [8], rhodanese [9], IFNγ [10]), or as high as 130°C (e.g. in thermophilic bacterial enzymes [11]). Given the temperature dependent effects on protein conformation, ideally, degradation kinetics should be studied as close to the temperature of interest as possible, and far away from T_m, so as to avoid additional degradation pathways (e.g. aggregation) not relevant to lower temperatures. How much below the T_m of a protein should be used to make reasonable predictions about processes, say at 5°C? This would be protein-specific, depending on the protein's T_m, its thermodynamic stability, and chemically reactive sites. For example, RNase with a T_m of 51.4°C, an unfolding enthalpy of 96.6 kcal/mol at the T_m, a heat capacity change accompanying unfolding of 12.5 cal/mol/deg would contain 0.005% and 0.3% unfolded species at 25°C and 35°C, respectively [12]. Though the current analytical tools are not sensitive to directly detect such changes, an order of magnitude increase in the fraction of unfolded protein could potentially facilitate further physical destabilization. This may explain the observation of a substantial amount of conformationally altered species, for example at 20-40°C below the T_m [6] or during heat-treatment of urokinase for viral inactivation [13]. The formation of species with irreversibly altered conformation is also implicated by the presence of soluble aggregates, again at temperatures that are 30-40°C below the melting transition [6, 14]. Additionally, a small change in temperature may affect protein's physical stability, such that the population of partially unfolded species increases;

the time history and environment of these species then determines their fate, either reversibly converting them back to native form, or proceeding to aggregation.

Protein denaturation may also occur below ambient temperatures, "cold denaturation" [3, 15-17], as a result of weakening in the hydrophobic effect (due to increased solvation of the hydrophobic residues). The weakening of such intramolecular forces at low temperatures could also induce dissociation of multimeric proteins, leading to their inactivation [3].

The complexity in interpreting temperature-dependent observations in proteins, therefore, lies not only in the presence of multiple chemically reactive centers, but to the extent to which chemical modifications are coupled to conformational changes. For a single chemical reaction pathway, the temperature-dependence generally follows Arhenius relationship, indicating that a reaction rate (k) is dependent on the product of the frequency of collision of the reactants (A) and the chance that the colliding molecules have sufficient energy (E_a) for the reaction to occur:

$$\ln k = - E_a / RT + A$$

Despite the complexities described above, several studies have shown linear Arhenius profiles, not only for chemical degradation in proteins [18, 19], but also for enzyme inactivation during thermal/cold denaturation [11, 20]. This suggests either a lack of change in the reaction mechanisms or protein conformation around the reactive species over the temperature range studied, or similarity in the activation energy for the different primary reaction mechanisms (see Table 1 for typical values); one also can not preclude the presence of many complex reactions, each having a small contribution to the overall rate constant. Interestingly, linear Arhenius profiles have also been seen when molecular motions have been substantially reduced, as in the solid dehydrated state [20] or even after transition from liquid to solid solution at subzero temperatures [11]. Moreover, such a relationship has held even when complex reaction kinetic models have been evoked [21, 22], suggesting that, in many cases, it is reasonable to extrapolate from the high temperature data (e.g. 40°C) to predict shelf life at low temperatures (e.g. 5°C).

pH Dependence of Protein Stability:

The hydronium ion concentration in solution controls the ionization state of charged amino acids in proteins, altering the local electrostatic field gradient, and ionic and

Table 1 - Ranges of Activation Energies for Typical Protein Reaction Pathways

Asn deamidation	~20-25 kcal/mol [23]
Hydrolysis/Proteolysis	~17-25 kcal/mol (e.g. 20-22 kcal/mol for L-succinimide and ~25-26 kcal/mol for D-succinimide [24])
Racemization	25-30 kcal/mol (e.g. for aryl-Gly [25])
Proline Isomerization	~15 kcal/mol [26]
Thiol/disulfide Scrambling	~14 kcal/mol [27]
Oxidation	8-15 kcal/mol
	12-18 kcal/mol for thiols [28]
Protein denaturation	5-10 kcal/mol [12]

Table 2 - Common Chemical Degradation Pathways in Proteins

Pathway:	Sites and Usual Conditions
Deamidation:	Asn-X (X=Gly or Ser) and/or flexible loops and/or correct peptide backbone n+1 nitrogen orientation at alkaline pH [33-36] or high phosphate and carbonate [35].
	C-terminal Asn [37]
Isomerization/ Racemization:	Asp-X (X=Gly or Ser) and/or flexible loops [38], acid and base catalyzed [25].
Cyclic Imide:	Asp-X (X=Gly or Ser) [39-41].
Cleavage:	Asp-X (X usually Pro) at acid pH [42, 43] X-Gly (X= Asp, Asn, Lys, Arg) or Gly-X (X=Gly, Ser) [35, 44-46].
	Proteolytic
Oxidation:	Thiols in Cysteine when exposed (by oxidants and at alkaline pH) [28, 47, 48].
	Met when exposed[49, 50], metals or near metal chelating amino acids, anti-oxidants or ETDA[51, 52], light [53], oxidants[54], glycation[55]; usually pH independent, unless in high phosphate [56]
	His & Trp by oxidants & light [57].
Pyroglutamate:	Gln-X at alkaline or acid pH [58].
Diketopiperizine:	N-terminal amine-Gly-X (X=Gly or Pro); base-catalyzed[59].
β-elimination:	Disulfide at high pH [60-62]
Crosslinking:	Thiols by oxidants and at high pH [48] Lysinoalanine by-product of β-elimination [46] Dityrosine by-product of oxidation [63, 64] Maillard Reaction of Lys with reducing sugars and subsequent oxidation [65].

dipole-dipole interactions. At extemes of pH (typically at pH values below ~4 and above ~10), where charge density increases substantially, protein's global conformation is disrupted by electrostatic repulsions; in some proteins, self-association of these denatured species are observed [10, 14, 29]. Even within the physiological range, pH changes may lead to dissociation of multimeric proteins, generation of new binding sites for ions of buffers and excipients, and loss of a cofactor or metal binding [30, 31]. Some proteins show sharp (within one unit) pH-induced changes in physical-biochemical properties related to protonation of a critical amino acid [30-32]; most others have broad pH-denaturation profiles, consistent with global electrostatic effects [30]. Thus, an early assessment of the pH-dependent conformational state would be valuable in narrowing the limits of this important formulation design parameter.

In addition to changes in the physical state of the protein, pH can markedly affect chemical modification rates (see reviews by Manning, 1989; Wang, 1988; Cleland, 1994; Shahrokh, 1997[66-69]). As listed in Table 2, the hydronium ion is involved directly in specific acid-catalyzed reactions, as in Asp-Pro cleavage [42], direct Asn-Gly hydrolysis [35], and succinimide formation at Asn-X residues [39], and under some conditions. Some specific-base (hydroxide ion) catalyzed reactions include succinimide hydrolysis & Asn deamidation [35], thiol crosslinking, disulfide scrambling and β-elimination [61], diketopiperazine [59] and pyroglutamic acid [58] product formation.

From the perspective of the protein's environment, pH also plays a critical role in maintaining glass container integrity [70], and leaching of extractables from rubber stoppers [71, 72]. Hence, alkaline pH promotes dissolution of silica, giving rise to particulates, and acid pH promotes metal leaching from rubber stoppers.

Chemical and Physical Degradation Pathways - Formulation Selectio Considerations:

The most common chemical degradation pathways in proteins are listed in Tabl (see reviews by Manning, 1989; Wang, 1993; Cleland, 1994; Pearlman, 19 Shahrokh, 1997[14, 29, 66, 68, 69]). Of these pathways, deamidation, disul crosslinking/scrambling, and pyroglutamate formation, occur readily above ne pH, and are enhanced in phosphate or bicarbonate buffers (see Table

Interestingly, the moderately alkaline pH and the phosphate-containing buffers that are typically used for purification and study of new biological entities, both facilitate such chemical reactions, and may lead to loss in biological activity [73]. The strategy of decreasing formulation pH minimizes these chemical reactions, but may promote cleavage at susceptible Asp-Pro motifs [35, 42, 43]. Thus, an optimal formulation should have a balanced choice of pH, buffer species, and ionic strength to maximally protect the protein's integrity.

Chemical reactivity is often coupled to protein conformation and the degree of surface exposure of the amino acids; in the case of deamidation, reactivity is also governed by the flexibility of the environment and the conformation of the C-terminal flanking residues [37, 74]. Hence, examination of the primary sequence and hydropathy plots provide a simplistic initial view of the potential "hot spots" on the protein. Local effects, such as selective Methionine oxidation due to proximity to metal chelating Histidine or Lysine residues [52], and increased susceptibility to deamidation of typically non-reactive (based on primary sequence motif) Asn residues [75-77] might be implicated by careful examination of the crystal structure. The influence of protein conformation on Asp-Pro cleavage has also been reported, in which native β-turn conformation promotes cleavage relative to the denatured form [78]. Ultimately, the empirical determination of the pH-profile of the chemical degradation rate constants provides the pH of optimal stability.

Table 3 lists some of the conditions that compromise the physical stability of proteins (see reviews by Ahern, 1992 [92]). Proteins are suceptible to loss of native conformation under conditions such as exposure to co-solvents or hydrophobic surfaces. Though reversible refolding (e.g., upon removal of the destabilizing condition) may occur (e.g. Dungan, 1993 [9]), the most common fate of the unfolded protein species in aqueous solutions is aggregation (to shield the exposed hydrophobic sites) and/or precipitation due to low solubility of the denatured form. In some proteins, sequence clusters of hydrophobic amino acids make them prone to aggregation [93]; in others, external conditions such as interfaces dictate the fate of the protein [86, 94]. For example, denaturation at air-water interfaces generally produces protein precipitates (see review by Horbett, 1992[95]). Also, adsorption to solid surfaces involves an initial "docking" of the native protein onto the surface, followed by flattening and denaturation [96]; this often leads to an irreversible loss of soluble protein to the surface. Interestingly, however, in a recent study of IL-2 delivery by Alzet minipumps, most of the adsorbed layer in the pump released into

Table 3 - Physical Degradation Pathways in Proteins

Pathway	Usual Conditions
Denaturation	Organic solvents (usually >10%); detergents (any ionic; >0.5 % non-ionic) Extremes of pH (~<4-5 and >9) and temperatures (~T_m, "cold denaturation"[3]) Adsorptive surfaces, delivery devices[79-83] Ice surface [84] Air-water interface [85, 86] High pressure [87] High ionic strength [88]
Denaturation/ Aggregation	Extremes of pH and temperatures Hydrophobic interfaces (air-water, ice-water, solid-water) Preservatives [89, 90] High protein concentration (usually >20-50 mg/ml) [91] Low protein concentration (e.g. <200 µg/ml for rhodanese) [9]
Dissociation	Low protein concentration Organic solvents, Chaotropes Temperature and pH extremes

the solution[79]. More intriguingly, the released protein, which was denatured, inactive, and at concentrations in the mg/ml range, remained monomeric rather than aggregated.

Excipients for prevention of protein aggregation seem to exert their effects through a number of pathways including: (a) binding to the protein and stabilizing the native conformation (e.g. polysulfates and FGFs [97, 98] or cyclodextrins [99]), (b) preferential exclusion from the protein surface, allowing for preferential hydration of the protein (e.g. PEGs and sugars [91, 100]), (c) prevention of protein-protein or protein-surface associations or increasing solubility of the aggregates (e.g., non-ionic surfactants (see reviews by Cleland, 1993; Brange, 1993 [101, 102])).

Solid State:

Hydrolytic, oxidative, or proteolytic degradation reactions compromise long term stability of many proteins in solution. A decrease in temperature, e.g. by freezing, would decrease chemical reactivity by decreasing molecular mobility. However, the protein's physical stability might be compromised by the rate and temperature of freezing [103]. As the temperature decreases below zero, ice crystals are formed and are phase-separated from a very concentrated solution of protein, salts, and other excipients including oxygen [104]. Slow freezing could prolong the exposure of the protein to a concentrated phase; very fast freezing yields small ice crystals with a large surface area that facilitate protein denaturation [103]. Selective excipient crystallization (e.g., mannitol and NaCl at -20°C) and the resulting pressure changes may also provide denaturing surfaces [84], or lead to drastic pH drops (e.g., Na phosphate [105]). An appropriate choice of buffer and inclusion of cryoprotectants (typically high molecular weight polymeric compounds such as polyethylene glycols) appear to bring freeze/thaw stability to many proteins. The mechanism of cryoprotection is believed to be exclusion of excipients from the protein surface and preservation of a hydration shell that maintains the native conformation via stabilizing compact native state over elongated denatured state[91, 100, 106, 107].

Further reduction of hydrolytic processes may be achieved by removal of water, by freeze-drying drying [84, 107-112] or spray-drying [113-116]. Strategic approaches to developing stable dry powder protein formulations have been the subject of extensive studies (see references just above and also Hageman, 1992[117]). The most prominant degradation pathway during this process is aggregation, due to exposure to ice surface (during freezing) or air surface (during drying), and freeze concentration of protein via excipient phase separation, all giving subsequent protein conformational loss [118, 119]. Solid-state FTIR studies have shown that excipients which minimize conformational changes during dehydration, provide long term storage stability [118]. These excipients appear to be sugars, believed to replace water and provide stabilization by hydrogen bonding to the protein. Also, inclusion of surfactants, albumin, or amino acids have been successful to minimize denaturation at surfaces [85, 116, 120] and particulate formation during storage of lyophilized proteins. An important consideration is the moisture content: as the moisture level increases, the chemical reactivity and physical stability increases, both by providing water as a reactant and by increasing molecular mobility [62, 103, 117, 121]. Hence, maintaining a low moisture content of ~2-4% has usually been optimal, the lower limit ensures maintenance of a water monolayer (0.25 g/g protein) around the protein surface for its conformational stability [103, 122]. To achieve this, consideration should also be made to the types of stoppers [123]. Moisture induced destabilization is a special concern in sustained release formulations, where the solid state protein in the polymeric matrix is exposed to the high moisture content of tissues (and accelerated temperature of the body) for long periods in vivo (see reviews by Cleland, 1994, Costantino, 1994, and Schwendenman, 1996 [62, 68, 124]).

Given the higher cost and time of developing dry powder formulations, liquid formulations would be the first choice if adequate protein stability is feasable. Special conditions, however, may dictate development of dry powder preparations, irrespective of solution stability. For example, solid preparations allow versatile reconstistution to achieve a range of high dose concentrations with only one vial configuration.

Analytical Tools for Stability Assessment:

Key to efficient formulation development is utilizing multiple, orthogonal, information-rich, and quantitative analytical tools. Rapid identification of primary degradation pathways facilitates rational optimization of formulations. Marketing a Biotechnology product in the 70's required demonstration of preserved biological activity (with the typical 30-50% assay variability), SDS-PAGE patterns, and visual clarity of the solution. Advances in technology now provide femtomolar detection of variants or impurities, and detailed information on carbohydrate sequence and conformation. The common analytical tools used for determining chemical and physical properties of proteins are listed in Table 4 (also see reviews by Jones, 1993, 1994[125, 126]).

Experimental Design Considerations:

Given the number of formulation parameters that can be altered, and the typically limited initial amount of material, strategic experimentation through the use of Experimental Design can be an efficient way to select a formulation, as has been effectively used in the chemical industry for many years [158-160]. Instead of changing one factor at-a-time, Experimental Design calls for using matrix methods that explores the experimental space by studying many variables at the same time in a systematic way, ultimately using less sample and time to reach the optimum (the details of this approach is beyond the scope of this review and the readers are referred to excellent books on the topic [160, 161]). For example, an initial screen of typical important formulation parameters such as pH, temperature, protein concentration (varied at 3-5 wide levels), plus 2-level parameters such as light (vs. dark), anti-oxidants (vs. none), tonicifier (sugar vs salt), surfactant (vs. none), etc., can rapidly sieve out critical parameters that affect protein stability and any potential interactive effects of those parameters on protein stability. In the second experiment, fewer critical factors are selected at levels nearer to the region of optimum, allowing one to build a predictive model, which is next verified at predicted best settings. The strength of this systematic approach lies in providing information-rich data that have statistical power through the use of averages. Examples of this approach that has been utilized in identifying optimal formulations and minimizing degradation in proteins are given by Massart, 1988, Fransson, 1996, Patel, 1990 [56, 162, 163].

Formulation Approaches and Future Considerations:

Efficient development of a stable formulation is critical for rapid initation of clinical evaluation of a therapeutic protein. Ideally, preformulation work should quickly reveal the bioactivity-compromising physical and chemical states of the protein. For example, given that neutralization of the overall charge at the isoelectric point gives minimal solubility of the native state, it is imperative to determine pI very early in preformulation studies. Also, assessment of the pH dependence of chaotrope- or temperature- induced denaturation could reveal the range of suitable pHs for optimal

Table 4 – Methods for Analysis of Physico-Chemical Status of Proteins

Parameter	Method (useful range)	Limitations	Information Content/Advantages
Size/ Molecular Association	Mass Spectrometry Electrospray (5-50 pmoles)	MW of ~150,000 Da; Dissociation of non-covalent aggregates; Have to minimize salt/detergent; Semi-quant. only for similar species.	Detects covalent aggregates & clips, glycosylation, post-translational modification, primary sequence; High mass accuracy (±0.01%); Can be on-line with LC.
	MALDI-TOF (5-1000 fmoles)	MW of ~200,000 Da; Matrix may affect detection; Have to minimize salt/detergent; Semi-quant. only for similar species.	Detects covalent aggregates & clips, sequence modification; Fast; Accurate.
	(Yates, 1996; Burlingame, 1996; Miranker, 1996; Nguyen, 1995)[127-130]		
	Size Exclusion HPLC (1-100 µg/species)	Broad peaks; Potential column interactions may distort MW estimation.	Estimates hydrodynamic diameter; Detects covalent & non-covalent aggregates & clips; Quantitative.

Table 4 - Methods for Analysis of Physico-Chemical Status of Proteins – Continued

Parameter	Method (useful range)	Limitations	Information Content/Advantages
	SDS-PAGE (2-10 μg/band)	Qualitative; Mobility affected by sialylated carbohydrates.	Crude MW estimation; Detection of covalent aggregates and clips.
	<u>Capillary Electrophoresis</u> CZE (0.05-0.2 μg) CE-SDS non-gel sieving (Righetti, 1996)[131]	Needs internal standards; Potential capillary interaction.	Detects size and charge variants; Quantitative; High resolution.
	<u>Light Scattering</u> (L.S.) (Harding, 1992; Arakawa, 1996)[132, 133]		
	Dynamic Laser L.S. (>0.1 mg/mL)	Over-sensitive to large species; Model-dependent association calc.; Prone to interference by non-protein.	Gives distribution of hydrodyn. radius (>3 nm); Can be on-line.
	(Mhatre, 1993; Kadima, 1993; Endo, 1992)[134-136]		
	Static Laser L.S.	Over-sensitive to large species; Model-dependent calculation	Gives average MW (>50 kDa), radius of gyration and association states; can be on-line.

continued on next page

Table 4 - Methods for Analysis of Physico-Chemical Status of Proteins – *Continued*

Parameter	Method (useful range)	Limitations	Information Content/Advantages
		of association states; Prone to interference by non-protein; Requires several protein concentrations or separation of species.	
	(Kunitani, 1997)[137]		
	UV scattering (>1 μg/ml)	Over-sensitive to large particles; Prone to interference by non-protein; Ambiguous dependence on size, shape, number.	Detects particles > 500 nm; Sensitive; easy.
	(Eberlein, 1994; Vrkljan, 1994; Eckhardt, 1991)[89, 138, 139]		
	Ultracentrifugation (>0.02 mg/ml at A214 nm)	Not sensitive to <5% aggregates; Model-dependent K_d determination; Potential pressure effects.	Gives average MW (~5% accuracy) from sed. equil.; Gives shape from sedimentation velocity; Gives associations (10^{-3} - 10^{-8} M) in solution; Quantitative; Simultaneous multi-sample analysis.
	(Yphantis, 1964; Harding, 1993)[140, 141]		

Table 4 - Methods for Analysis of Physico-Chemical Status of Proteins – *Continued*

Parameter	Method (useful range)	Limitations	Information Content/Advantages
	Calorimetry		
	Isothermal (>0.1 mg/mL)	Model-dependent K_d determination.	Gives binding constants (10^{-6} - 10^{-12} M) in solution.
	(Koenigbauer, 1992; Koenigbauer, 1994)[142, 143]		
	Differential Scanning (>0.5 mg/ml)	Hard to interpret peak shapes. Heating may influence protein/protein or protein/excipient interactions.	Gives phase transition temperature and thermodynamic parameters; Detects aggregation (non-quant.); Multi-component analysis may suggest domain folding.
	(Haun, 1995; Freire, 1995)[7, 144]		
	Surface Plasmon Resonance (micro-molar)	Potential immobilization artifacts; One ligand must be stable to multiple regenerations. Diffusional limitations of ligand can influence K_d values.	Gives binding constants (10^{-6} - 10^{-12} M) at surface/solution; Gives on/off rates
	(Hutchinson, 1995; Malmqvist, 1994; Garland, 1996)[145-147]		

continued on next page

Table 4 -Methods for Analysis of Physico-Chemical Status of Proteins – *Continued*

Parameter	Method (useful range)	Limitations	Information Content/Advantages
Chemical Heterogeneity	HPLC (Reverse phase, Ion Exchange, Size Exclusion; Hydrophobic interaction, Hydrophilic interaction) Potential recovery loss (HIC).		Quantitative; Sensitive (detects 0.1% variants).
Charge	Isoelectric Focusing (native) (~ 2-10 µg/band)	Qualitative; Can be smeary for glycoproteins; Protein may precipitate near pI.	Shows heterogeneity at native conditions which may be conformation-dependent.
	Urea-IEF (denatured)	Carbamoylation by urea.	Gives conformation-independent pI; Urea prevents precipitation during focusing.
	(Walker, 1994; Gianazza, 1995)[148,149]		
	CE-IEF (0.05-0.2 µg)	Need internal markers due to run-to-run mobility variations; May need modifiers to keep protein in solution near pI.	Quantitative; High resolution; Native conditions
	(Liu, 1996; Righetti, 1996)[131,150]		
	Ion exchange HPLC (1-100 µg/species)	Broad peaks.	Quantitative; Native conditions; May separate similar charge species with different polarities (e.g., deamidation at different sites)

Table 4 - Methods for Analysis of Physico-Chemical Status of Proteins – *Continued*

Parameter	Method (useful range)	Limitations	Information Content/Advantages
	Peptide Mapping (~1 μg/peak)	Labor-intensive; Potential selective peak recovery loss; Usually overlapping peaks; Non-specific cleavages dependent on trypsin batch.	Easier to characterize than whole molecule; Sensitive identity test; Can be automated; Can do conformational mapping of native protein; Can also detect impurities.
	(Dong, 1992; Shieh, 1995; Billeci, 1993; Dogruel, 1995)[151-154]		
	Sequencing (>20 picomoles)	Methionine-sulfoxide unstable.	Identity test; Shows heterogeneity; Quantitative; Stops at Iso-asp or succinimide
	Amino Acid Analysis (1 μg proteins, 100 pmol peptides)		Gives concentration, amino acid modifications;
Conformation (Spectroscopic tools)	Circular Dichroism (>0.1 mg/mL)	Subject to interference with non-protein materials; Model-dependent fits of structure (±20%).	Detects secondary structure (far UV), tertiary structure and S-S (near UV).
	(Havel, 1996)[155]		

continued on next page

Table 4 -Methods for Analysis of Physico-Chemical Status of Proteins – *Continued*

Parameter	Method (useful range)	Limitations	Information Content/Advantages
	Fluorescence (1-100 μM)	Source of change in signal may be ambiguous.	Detects tryptophan/tyrosine environment (tertiary structure), Folding/unfolding, Molecular interactions, Ligand binding, Hydrodynamic radius (from polarization), Intermolecular distance from energy-transfer studies), Time-resolved (ms) kinetics of diffusion, & association; Can study solid-state or turbid solutions.
	(Jiskoot, 1995)[156]		
	Vibrational (Raman, FTIR) (>0.02% w/v μM to mM)	Model-dependent fit for 2° struc.; Major water signal interference.	Detects carbonyl stretch; Can interpolate secondary structure ; Quantitative; Can study solid state.
	(vanStokkum, 1995; Hendra, 1996; Havel, 1996)[6, 155, 157]		
Structure	X-ray crystallography, NMR (solution and solid-state)		

conformational stability [164]. Such studies can be designed to identify excipients/conditions that stabilize the native conformation. Other stress conditions can be used to test the physical stability of the protein and to screen excipients that would minimize its alteration. For example, a simple study of agitation-induced aggregation/precipitation may point to the need for inclusion of non-ionic surfactants, albumin, sugars, or amino acids. Also, identifying the need for and the level of surfactant to prevent potential adsorption to surfaces could substantially improve handling. Finally, probing of the protein's chemical stability, e.g. susceptibility to peroxides [49], could identify the bioactivity-compromising reactive sites, and guide formulation development.

It is clear that efficient and economic approaches to test the stability of protein formulations are highly desirable. Several new information-rich, sensitive, quantitative, and fast technologies have been developed which facilitate this goal. For example, CE-IEF provides quantitative, faster, more sensitive information on charge heterogeneity compared to the conventional IEF method [131]. Orthogonal, coupled HPLC techniques may efficiently identify peaks corresponding to different stability indicating assays, allowing discontinuation of redundant assays earlier in the development phase. Other coupled technologies, such as LC/MS/MS, efficiently provide sequence information on multiple peaks in peptide maps, and considerably accelerate degradation product identification. Information from SEC may be enriched through the use of an on-line light scattering detector for sensitive aggregate quantitation [135, 136, 165] and for MW distribution across peaks [166]. Also, on-line fluorescence detectors allow simultaneous detection of native and denatured species [165]. Techniques such as fluorescent staining of SDS-PAGE gels give quantitative measure of MW distributions, with similar sensitivity as silver staining, and markedly greater convenience [167]. MALDI analysis of mixtures may give a more accurate and sensitive preview of the distribution of covalent aggregates and clips than conventional SDS-PAGE [126]. Finally, quantitative peptide mapping is being developed as a high resolution information-rich assay of protein identity and stability [168-170].

A limitation in developing optimal protein formulations is lack of models that can predict immunogenecity of degradation components, since animal models are not predictive of humans. Moreover, novel and safe excipients for improved long term stability of proteins are rather limited. Additionally, non-parenteral routes have not generally been successful delivery means, creating major challenges in long term therapy. Future efforts should be directed at identifying approaches for facile localized administration, minimizing the drug load while maximizing patient compliance.

Acknowledgments: I greatly appreciate discussions and critical input by Drs. Linda De Young, Jun Liu, Reed Harris, and Jeff Cleland at Genentech Inc.

Literature Cited:

1) F. W. Anderson. Human Gene Ther. **1995**; 6:275-276.
2) C. B. Anfinsen. Biochem. J. **1972**; 128:737-749.
3) R. Jaenicke. Phil. Trans. Royal Soc. London - Series B: Biological Sci. **1990**; 326:535-551.
4) D. Bashford. Curr. Opin. Struct. Biol. **1991**; 1:175-184.
5) P. L. Privalov and S. J. Gill. Adv. Protein Chem. **1988**; 39:191-234.
6) I. H. M. vanStokkum, H. Linsdell, J. M. Hadden, P. I. Haris, D. Chapman and M. Bloemendal. Biochemistry **1995**; 34:10508-10518.
7) M. F. Haun, M. Wirth and H. Ruterjans. Eur. J. Biochem. **1995**; 227:516-523.
8) R. A. Copeland, H. Ji, A. J. Halfpenny, R. W. Williams, K. C. Thomspon, W. K. Herber, K. A. Thomas, M. W. Bruner, J. A. Ryan, D. Marquis-Omer, G. Sanyal, R. D. Sitrin, S. Yamazaki and C. R. Middaugh. Arch. Biochem. Biophys. **1991**; 289:53-61.
9) J. M. Dungan and P. M. Horowitz. J. Protein Chemistry **1993**; 12:311-321.
10) M. G. Mulkerrin and R. Wetzel. Biochemistry **1989**; 28:6556-6561.
11) N. More, R. M. Daniel and H. H. Petach. Biochem. J. **1995**; 305:17-20.
12) C. N. Pace, B. A. Shirley and J. A. Thomson. In: Protein Structure. . Oxford, England: IRL Press, 1989:311.
13) W. R. Porter, H. Staack, K. Brandt and M. C. Manning. Thrombosis Res. **1993**; 71:265-279.
14) Y. J. Wang and R. Pearlman, *Stability and characterization of protein and peptide drugs*. Pharmaceutical Biotechnology, ed. Vol. 5. 1993, New York: Plenum Press.
15) P. L. Privalov. CRC Crit. Rev. Biochem. Mol. Biol. **1990**; 25:281-305.
16) D. T. Brandau, E. Q. Lawson, C. R. Middaugh and G. W. Litman. Immunol. Invest. **1986**; 15:447-462.
17) K. E. Hightower and R. E. McCarty. Biochemistry **1996**; 35:4852-4857.
18) V. J. Helm and B. W. Muller. Pharm. Res. **1990**; 7:1253-1256.

19) J. L. Wolfe, G. E. Lee, G. K. Potti and J. F. Gallelli. J. Pharm. Sci. **1994**; 83:1762-1764.

20) S. Yoshioka, Y. Aso, K.-I. Izutsu and T. Terao. J. Pharm. Sci. **1994**; 83:454-456.

21) S. Yoshioka, K.-I. Izutsu, Y. Aso and Y. Takeda. Pharm. Res. **1991**; 8:480-484.

22) L. C. Gu, E. A. Erdos, H.-S. Chiang, T. Calderwood, K. Tsai, G. C. Visor, J. Duffy, W.-C. Hsu and L. C. Foster. Pharm. Res. **1991**; 8:485-490.

23) K. Patel and R. T. Borchardt. Pharm. Res. **1990**; 7:703-711.

24) T. Geiger and S. Clarke. J. Biol. Chem. **1987**; 262:785-794.

25) G. G. Smith and T. Sivakua. J. Organic Chem. **1983**; 48.

26) J. F. Brandts, H. R. Halvorson and M. Brennan. Biochemistry **1975**; 14:4953-4963.

27) N. Darby and T. E. Creighton. In: Methods in Molecular Biology. B. A. Shirley, ed. Vol. 40. Totowa NJ: Humana Press Inc., 1995:219-223.

28) T. H. Nguyen. In: Formulation and Delivery of Proteins and Peptides. J. L. Cleland and R. Langer, eds. . Washington DC: American Chemical Society, 1994:59-71.

29) R. Pearlman and Y. J. Wang, *Formulation, Characterization, and Stability of Protein Drugs: Case Histories*. Pharmaceutical Biotechnology, ed. Vol. 9. 1996, New York: Plenum Press.

30) J. B. Matthew, F. R. N. Gurd, B. Garcia-Moreno, M. A. Flanagan, K. L. March and S. J. Shire. CRC Crit. Rev. Biochem. **1985**; 18:91-197.

31) A. Artigues, A. Iriarte and M. Martinez-Carrion. J. Biol. Chem. **1994**; 269:21990-21999.

32) J. T. Yang. Tanpakushitsu Kakusan Koso - Protein, Nucleic Acid, Enzyme. **1981**; 26:803-814.

33) A. B. Robinson and C. J. Rudd. Curr. Topics Cell. Regulation **1974**; 8:247-295.

34) N. V. Pushikina, I. E. Tsybul'skii and A. I. Lukash. Prikladnaia Biokhimiia Mikrobiol. **1986**; 22:198-204.

35) K. Patel and R. T. Borchardt. J. Parenteral Sci. Tech. **1990**; 44:300-301.

36) C. Oliyai and R. T. Borchardt. Pharm. Res. **1993**; 10:95.

37) S. Clarke, R. C. Stephenson and J. D. Lowenson. In: Stability of Protein Pharmaceuticals. Part A: Chemical and Pathways of Protein Degradation. T. J. Ahern and M. C. Manning, eds. . New York: Plenum Press, 1992:1.

38) J. L. Radkiewicz, H. Zipse, S. Clarke and K. N. Houk. J. Am. Chem. Soc. **1996**; 118:9148-9155.

39) B. N. Violand, M. R. Schlittler, E. W. Kolodziej, P. C. Toren, M. A. Cabonce, N. R. Siegel, K. L. Duffin, J. F. Zobel, C. E. Smith and J. S. Tou. Protein Science **1992**; 1:1634-1641.

40) R. I. Senderoff, S. C. Wootton, A. M. Boctor, T. M. Chen, A. B. Giordani, T. N. Julian and G. W. Radebaugh. Pharm. Res. **1994**; 11:1712-1720.

41) Z. Shahrokh, G. Eberlein, D. Buckley, M. V. Paranandi, D. W. Aswad, P. Stratton, R. Mischak and Y. J. Wang. Pharm. Res. **1994**; 11:936-944.

42) D. Piszkiewica, M. Landon and E. L. Smith. Biochem. Biophys. Res. Commun. **1970**; 40:1173-1178.

43) T. J. Ahern and A. M. Klibanov. Methods Biochem. Anal. **1988**; 33:91-127.

44) S. E. Zale and A. M. Klibanov. Biochem. **1986**; 25:5432-5444.

45) J. D. Kowit and J. Maloney. Anal. Biochem. **1982**; 123:86-93.

46) J. J. Correia, L. D. Lipscomb and S. Lobert. Arch. Biochem. Biophys. **1993**; 300:105-114.

47) T. E. Creighton, A. Zapun and N. J. Darby. Trends In Biotechnology **1995**; 13:18-23.

48) T. Kortemme and T. E. Creighton. J. Mol. Biol. **1995**; 253:799-812.

49) R. G. Keck. Anal. Biochem. **1996**; 236:56-62.

50) V. Mihajlovic, O. Cascone and M. J. Biscoglio-de-Jimenez-Bonino. Intl. J. Biochem. **1993**; 25:1189-1193.

51) S. Li, T. H. Nguyen, C. Schoneich and R. T. Borchardt. Biochemistry **1995**; 34:5762-5772.

52) S. Li, C. Schoneich and R. T. Borchardt. Pharm. Res. **1995**; 12:348-355.

53) D. C. Cipolla and S. J. Shire. In: Techniques in Protein Chemistry II. J. J. Villafranca, ed. . San Diego: Academic Press, 1991:543-555.

54) T. H. Nguyen, J. Burnier and W. Meng. Pharm. Res. **1993**; 10:1563-1571.

55) P. K. Hall and R. C. Roberts. Biochimica Biophys. Acta **1992**; 1121:325-330.

56) J. Fransson and A. Hagman. Pharm. Res. **1996**; 13:1476-1481.

57) E. R. Stadtman. Ann. Rev. Biochem. **1993**; 62:797-821.

58) K. M. Khandke, T. Fairwell, B. T. Chait and B. N. Manjula. Intl. J. Pept. Prot. Res. **1989**; 34:118-123.

59) N. F. Sepetov, M. A. Krymsky, M. V. Ovchinnikov, Z. D. Bespalova, O. L. Isakova, M. Soucek and M. Lebl. Peptide Research **1991**; 4:308-313.

60) J. R. Whitaker and R. E. Feeney. Critical Reviews in Food Science & Nutrition **1983**; 19:173-212.

61) D. B. Volkin and C. R. Middaugh. In: Stability of Protein Pharmaceuticals. Part A: Chemical and Physical Pathways of Protein Degradation. T. J. Ahern and M. C. Manning, eds. . New York: Plenum Press, 1992:215.

62) H. R. Costantino, R. Langer and A. M. Klibanov. J. Pharm. Sci. **1994**; 83:1662-1669.

63) D. A. Malencik and S. R. Anderson. Biochemistry **1996**; 35:4375-4386.

64) C. Giulivi and K. J. Davies. Methods in Enzymolgy **1994**; 233:363-371.

65) M. X. Fu, K. J. Wells-Knecht, J. A. Blackledge, T. J. Lyons, S. R. Thorpe and J. W. Baynes. Diabetes **1994**; 43:676-683.

66) M. C. Manning, K. Patel and R. T. Borchardt. Pharm. Res. **1989**; 6:903-918.

67) Y. J. Wang and M. A. Hanson. J. Parenteral Sci. Tech. **1988**; 42:S4-S24.

68) J. L. Cleland and R. Langer. In: Formulation and Delivery of Proteins and Peptides. J. L. Cleland and R. Langer, eds. Vol. 567. Washington DC: American Chemical Society, 1994.

69) Z. Shahrokh and M. Powell. In: Peptide and Protein Drug Delivery. V. Lee, ed. : Marcel Dekker, 1997.

70) P. B. Adams. Bull. Parenteral Drug Assoc. **1977**; 31:213-226.

71) J. B. Boyett and K. E. Avis. Bull. Parenteral Drug Assoc. **1975**; 29:1-17.

72) S. Li, C. Schoneich, G. S. Wilson and R. T. Borchardt. Pharm. Res. **1993**; 10:1572-1579.

73) J. H. McKerrow and A. B. Robinson. Anal. Biochem. **1971**; 42:565-568.

74) M. F. Powell, G. Amphlett, J. Cacia, W. Callahan, E. Canova-Davis, B. Chang, J. L. Cleland, T. Darrington, L. DeYoung, B. Dhingra, R. Everett, L. Foster, J. Frenz, A. Garcia, D. Giltinan, G. Gitlin, W. Gombotz, M. Hageman, R. Harris, D. Heller, A. Herman, S. Hershenson, M. Hora, R. Ingram, S. Janes, M. Kamat, D. Kroon, R. G. Keck, E. Luedke, L. Maneri, C. March, L. McCrossin, T. Nguyen, S. Patel, H. Qi, M. Rohde, B. Rosenblatt, N. Sahakian, Z. Shahrokh, S. Shire, C. Stevenson, K. Stoney, S. Thompson, G. Tolman, D. Volkin, Y. J. Wang, N. Warne and C. Watanabe. In: Formulation, Characterization, and Stability of Protein Drugs. R. Pearlman and Y. J. Wang, eds. . New York: Plenum Press, 1996:1-140.

75) A. A. Kossiakoff. Science **1988**; 240:191-94.

76) R. Lura and V. Schirch. Biochemistry **1988**; 27:7671-7677.

77) C. L. Stevenson, A. R. Friedman, T. M. Kubiak, M. E. Donlan and R. T. Borchardt. Intl. J. Peptide Protein Res. **1993**; 42:497-503.

78) I. Segalas, R. Thai, R. Menez and C. Vita. FEBS Letters **1995**; 371:171.

79) S. T. Tzannis, W. J. M. Hrushesky, P. A. Wood and T. M. Przybycien. Proc. Natl. Acad. Sci. **1996**; 93:5460-5465.

80) L. V. Allen-Jr. PDA J. Pharm. Sci. Tech. **1995**; 49:306-308.

81) L. A. Trissel. PDA J. Pharm. Sci. Tech. **1995**; 49:309-313.

82) E. Changelau, G. Lange, M. Gasthaus, M. Boxberger and M. Berger. Diabetes Care **1987**; 10:348-351.

83) J. D. Andrade, V. Hlady, A.-P. Wei, C.-H. Ho, A. S. Lea, S. I. Jeon, Y. S. Lin and E. Stroup. Clin. Mater. **1992**; 11:67.

84) K. Izutsu, S. Yoshioka and S. Kojima. Pharm. Res. **1994**; 11:995-999.

85) M. Katakam, L. N. Bell and A. K. Banga. J. Pharm. Sci. **1995**; 84:713-716.

86) S. A. Charman, K. L. Mason and W. N. Charman. Pharm. Res. **1993**; 10:954-962.

87) J. L. Silva and G. Weber. Ann. Rev. Phys. Chem. **1993**; 44:89-113.

88) A. S. Yang and B. Honig. J. Molecular Biol. **1994**; 237:602-614.

89) G. A. Eberlein, P. R. Stratton and Y. J. Wang. PDA J. Pharm. Sci. Tech. **1994**; 48:224-230.

90) X. M. Lam, T. W. Patapoff and T. H. Nguyen. Pharm. Res. **1997**; in press.

91) T. Arakawa, Y. Kita and J. F. Carpenter. Pharm. Res. **1991**; 8:285-291.

92) T. J. Ahern and M. C. Manning, *Stability of Protein Pharmaceuticals. Part A. Chemical and Physical Pathways of Protein Degradation.* Pharmaceutical Biotechnology, ed. Vol. 2. 1992, New York: Plenum Press.

93) D. Burdick, B. Soreghan, M. Kwon, J. Kosmoski, M. Knauer, A. Henschen, J. Yates, C. Cotman and C. Glabe. J. Biol. Chem. **1992**; 267:546-554.

94) W. Norde. Adv. Colloid Interface Sci. **1986**; 25.

95) T. A. Horbett. In: Stability of Protein Pharmaceuticals. Part A. Chemical and Physical Pathways of Protein Degradation. T. J. Ahern and M. C. Manning, eds. Vol. 2. New York: Plenum Press, 1992:195-214.

96) A. Sadana. Chem. Rev. **1992**; 92:1799-1818.

97) D. B. Volkin and C. R. Middaugh. In: Formulation, Characterization, and Stability of Protein Drugs. R. Pearlman and Y. J. Wang, eds. . New York: Plenum Press, 1996:181.

98) Y. J. Wang, Z. Shahrokh, S. Vemuri, G. Eberlein, I. Beylin and M. Busch. In: Formulation. Characterization, and Stability of Protein Drugs: Case Histories. R. Pearlman and Y. J. Wang, eds. Vol. 9. New York: Plenum Press, 1996:141-.

99) M. E. Brewster, M. S. Hora, J. W. simpkins and N. Bodor. Pharm. Res. **1991**; 8:792-795.

100) S. N. Timasheff and T. Arakawa. In: Protein Structure, A Practical Approach. T. E. Creighton, ed. . New York: IRL Press, 1989:331-345.

101) J. L. Cleland, M. F. Powell and S. J. Shire. Crit. Rev. Ther. Drug Car. Sys. **1993**; 10:307-377.
102) J. Brange and L. Langkjaer. In: Stability and Characterization of Protein and Peptide Drugs: Case Histories. Y. J. Wang and R. Pearlman, eds. . New York: Plenum Press, 1993:315.
103) C. C. Hsu, H. M. Nguyen, D. A. Yeung, D. A. Brooks, G. S. Koe, T. A. Bewley and R. Pearlman. Pharm. Res. **1995**; 12:69-77.
104) S. Schwimmer, *Source Book of Food Enzymology*. 1981, Westport, CN: Avi Publishing.
105) F. Franks. Cryoletters **1990**; 11:93-110.
106) T. Arakawa and S. N. Timasheff. Biochemistry **1985**; 24:6756-6762.
107) J. F. Carpenter and J. H. Crowe. Cryobiology **1988**; 25:244-255.
108) M. J. Pikal, K. M. Dellerman, M. L. Roy and R. M. Riggin. Pharm. Res. **1991**; 8:427-436.
109) K. Inazu and K. Shima. Dev. Biol. Standardization **1992**; 74:307-322.
110) M. P. teBooy, R. A. deRuiter and A. L. deMeere. Pharm. Res. **1992**; 9:109-114.
111) J. F. Carpenter, T. Arakawa and J. H. Crowe. Dev. Biol. Standardization **1992**; 74:225-238.
112) B. S. Chang and N. L. Fischer. Pharm. Res. **1995**; 12:831-837.
113) J. Broadhead, S. K. E. Rouan, I. Hau and C. T. Rhodes. J. Pharm. Pharmacol. **1994**; 46:458-467.
114) M. Mumenthaler, C. C. Hsu and R. Pearlman. Pharm. Res. **1994**; 11:12.
115) T. P. Foster and M. W. Leatherman. Drug. Dev. Ind. Pharmacy **1995**; 21:1705-1723.
116) Y.-F. Maa, H. R. Costantino, P.-A. Nguyen and C. C. Hsu. Pharm. Res. **1997**.
117) M. J. Hageman. In: Stability of Protein Pharmaceuticals. Part A: Chemical and Physical Pathways of Protein Degradation. T. J. Ahern and C. M. M, eds. . New York: Plenum Press, 1992:273.
118) S. J. Prestrelski, N. Tedeschi, T. Arakawa and J. F. Carpenter. Biophys. J. **1993**; 65:661-671.
119) A. Dong, S. J. Prestrelski, S. D. Allison and J. F. Carpenter. J. Pharm. Sci. **1995**; 84:415-424.
120) S. Vemuri, C. D. Yu and N. Roosdorp. PDA J. Pharm. Sci. Tech. **1994**; 48:241-246.
121) D. Grieff. Cryobiology **1971**; 8:145-152.

122) J. H. Crowe, J. F. Carpenter, L. M. Crowe and T. J. Anchordoguy. Cryobiology **1990**; 27:219-231.
123) F. DeGrazio and K. Flynn. J. Parenteral Sci. Tech. **1992**; 46:54-61.
124) S. P. Schwendeman, M. Cardamone, M. R. Brandon, A. Klibanov and R. Langer. In: Microspheres/Microparticles-Characterization and Pharmaceutical Application. S. Cohen and H. Bernstein, eds. . New York: Marcel Dekker, Inc., 1996:1-49.
125) A. J. S. Jones. Adv. Drug Del. Rev. **1993**; 10:29-90.
126) A. J. S. Jones. In: Formulation and Delivery of Proteins and Peptides. J. L. Cleland and R. Langer, eds. . Washington DC: American Chemical Society, 1994:22-45.
127) J. R. Yates. Methods in Enzymology **1996**; 271:351-377.
128) A. L. Burlingame. Curr. Opin. Biotech. **1996**; 7:4-10.
129) A. Miranker, C. V. Robinson, S. E. Radford and C. M. Dobson. FASEB J. **1996**; 10:93-101.
130) D. N. Nguyen, G. W. Becker and R. M. Riggin. J. Chromatography **1995**; 705:21-45.
131) P. G. Righetti, *Capillary Electrophoresis in Analytical Biotechnology*. 1996, Boca Raton: CRC Press.
132) S. E. Harding, D. B. Sattelle and V. A. Bloomfield, *Laser Light Scattering in Biochemistry*. 1992, Cambridge: Royal Soc. of Chemistry.
133) T. Arakawa. Cell Tech. **1996**; 15:679-688.
134) R. Mhatre and I. S. Krull. Anal. Chem. **1993**; 65:283-286.
135) W. Kadima, L. Ogendal, R. Bauer, N. Kaarsholm, K. Brodersen, J. F. Hansen and P. Porting. Biopolymers **1993**; 33:1643-1657.
136) Y. Endo, H. Nagai, Y. Watanabe, K. Ochi and T. Takagi. J. Biochem. **1992**; 112:700-706.
137) M. Kunitani, S. wolfe, S. Rama and G. Dollinger. *1st Symposium on the Analysis of Well Characterized Biotechnology Pharmaceuticals*. 1997. San Francisco.
138) M. Vrkljan, T. M. Foster, M. E. Powers, J. Henkin, W. R. Porter, H. Staack, J. F. Carpenter and M. C. Manning. Pharm. Res. **1994**; 11:1004-1008.
139) B. M. Eckhardt, J. Q. Oeswein and T. A. Bewley. Pharm. Res. **1991**; 8:1360-1364.
140) D. A. Yphantis. Biochemistry **1964**; 3:297-317.
141) S. E. Harding. Biotech. Genet. Eng. Rev. **1993**; 11:317-356.

142) M. J. Koenigbauer, S. H. Brooks, G. Rullo and R. A. Couch. Pharm. Res. **1992**; 9:939-944.

143) M. J. Koenigbauer. Pharm. Res. **1994**; 11:777-783.

144) E. Freire. Methods in Molecular Biol. **1995**; 40:191-218.

145) A. M. Hutchinson. Molecular Biotech. **1995**; 3:47-54.

146) M. Malmqvist. J. Molecular Recognition **1994**; 7:1-7.

147) P. B. Garland. Qtrly. Rev. Biophys. **1996**; 29:91-117.

148) J. M. Walker. Methods Molecular Biol. **1994**; 32:59-65.

149) E. Gianazza. J. Chromatography **1995**; 705:67-87.

150) X. Liu, Z. Sosic and I. S. Krull. J. Chromatography **1996**; 735:165-190.

151) M. W. Dong. Adv. Chromatography **1992**; 32:21-51.

152) H. M. Shieh, R. T. Bass, B. S. Wang, M. J. Corbett and B. L. Buckwalter. J. Endocrinology **1995**; 145:167-174.

153) T. M. Billeci and J. T. Stults. Anal. Chem. **1993**; 65:1709-1716.

154) D. Dogruel, P. Williams and R. W. Nelson. Anal. Chem. **1995**; 67:4343-4348.

155) H. A. Havel, *Spectroscopic Methods For Determining Protein Structure in Solution*. 1996, New York: VCH Publishers. 1-246.

156) W. Jiskoot, V. Hlady, J. J. Naleway and J. N. Herron. In: Physical Methods to Characterize Pharmaceutical Proteins. J. N. Herron, W. Jiskoot and D. J. A. Crommelin, eds. Vol. 7. New York: Plenum Press, 1995:1-53.

157) P. J. Hendra. Amer. Lab. **1996**:17-24.

158) O. L. Davies, *Design and Analysis of Industrial Experiments*. 1956, New York: Hafner Publishing Company.

159) G. Taguchi, *Introduction to Quality Engineering: Designing Quality into Products and Processes*. 1986, Tokyo, Japan: Asian Productivity Organization.

160) M. D. Boleda, P. Briones, J. Farres, L. Tyfield and R. Pi. BioTechniques **1996**; 21:134-140.

161) S. Bolton, *Pharmaceutical Statistics: Practical and Clinical Applications*. 1984, New York: Marcel Dekker, Inc.

162) D. L. Massart and L. Buydens. J. Pharm. Biomed. Anal. **1988**; 6:535-545.

163) J. P. Patel, K. Marsh, L. Carr and G. Nequist. Intl. J. Pharmaceutics **1990**; 65:195-200.

164) R. Khurana, A. T. Hate, U. Nath and J. B. Udgaonkar. Protein Science **1995**; 4:1133-1144.

165) V. Sluzky, S. Z., P. Stratton, G. Eberlein and Y. J. Wang. Pharm. Res. **1994**; 11:485.

166) A. C. Herman. *1st Symposium on the Analysis of Well Characterized Biotechnology Pharmaceuticals*. 1997. San Francisco.

167) T. H. Steinberg, L. J. Jones, R. P. Haugland and V. L. Singer. Anal. Biochem. **1996**; 239:223-237.

168) A. Lim, E. Canova-Davis, V. Ling, M. Eng, L. Truong, B. Henzel, J. Stults, R. Harris, J. Gorrell, H. Heinsohn, C. McHugh, R. P. Weissburg and M. F. Powell. *Western Regional AAPS*. 1996. SSF, CA.

169) S. Renlund, I. M. Lintrot, M. Nunn, J. L. Schrimsher, C. Wernstedt and U. Hellman. J. Chromatography **1990**; 512:325-335.

170) S. A. Charman, L. E. McCrossin and W. N. Charman. Pharm. Res. **1993**; 10:1471-1477.

Chapter 2

The Pharmaceutical Development of Insulin: Historical Perspectives and Future Directions

Henry R. Costantino[1,3], Stanley Liauw[1], Samir Mitragotri[1], Robert Langer[1], Alexander M. Klibanov[2], and Victoria Sluzky[1,4]

Departments of [1]Chemical Engineering and [2]Chemistry, Massachusetts Institute of Technology, Cambridge, MA 02139

The pharmaceutical development of insulin has revolutionized diabetes therapy since it was introduced and clinically validated in the 1920's. Due to its importance for therapeutic use in a large and expanding market, insulin has been the focus of intense research in both academia and industry. As a result, there is currently much information available on the physical, chemical and biological aspects of insulin activity. Over the course of the clinical advancement of insulin, there has been a corresponding development in other areas related to biopharmaceutics, e.g., in federal regulation of biological drugs. Insulin formulations have advanced since the early days of crude pancreatic extracts to complex biphasic mixtures involving highly purified preparations of genetically engineered recombinant human protein. Insulin still remains a model for the biopharmaceutical industry. Current research in insulin therapy for diabetes is aimed at improving the insulin molecule and its formulations, developing novel methodologies for its delivery, and further understanding its stability.

Despite some 70 years of clinical use, the polypeptide hormone insulin is still undergoing new developments for application in the treatment of diabetes mellitus. From the early years of insulin's discovery to date, pharmaceutical investigators have competed to discover new, efficacious insulin formulations to bring to the market. Initial efforts focused on producing more pure insulin for rapid and prolonged actions in more stable formulations. As the protein's use became widespread in the U. S. and elsewhere, inconsistencies in insulin quality necessitated the development of federal regulation of biological drugs. In addition, the pharmaceutical development of insulin paved the way for numerous scientific advances, such as in protein chemistry and molecular biology. Due to its medical importance, insulin has been intensely studied and was one of the first proteins to be sequenced, examined by X-ray crystallography,

[3]Current address: Department of Pharmaceutical Research and Development, Genentech, Inc., 460 Point San Bruno Boulevard, South San Francisco, CA 94080

[4]Current address: COR Therapeutics, Inc., 256 East Grand Avenue, South San Francisco, CA 94080

synthesized, and later produced via recombinant DNA technology. Ultimately, insulin became the first recombinant human protein to become commercialized in the early 1980's. Today, it remains at the forefront in the quest for new technologies for polypeptide delivery, and is often employed as a model for investigation of protein stability. The story of the pharmaceutical development of insulin may shed light on the future of current investigational and newly discovered biopharmaceuticals.

The Polypeptide Hormone Insulin. Insulin is a peptide hormone involved in the regulation of glucose transport from the blood into cells which metabolize glucose (1). Insulin induces changes in the permeability of plasma membranes which in turn increase the uptake of glucose, along with various ions and biomolecules present in the blood. These effects lead to the overall anabolic effect of insulin including not only the metabolism of glucose, but also an increase in synthesis of glycogen, lipids and certain proteins. For a recent review on the physiologic basis of insulin, and its pharmacokinetics and pharmacodynamics in the treatment of diabetes, see Chien (2).

Because of its pharmaceutical importance, insulin's chemical nature has been the subject of intensive study, as reviewed by Brange and Langkjær (3). Briefly, the insulin monomer is composed of two amino-acid chains (A and B) and contains one intrachain and two interchain disulfide bonds (Figure 1). The biologically active form of insulin which binds to the insulin receptor is monomeric. Insulin monomers associate into dimers in a stable configuration that buries part of the hydrophobic surface of the monomer. In the presence of divalent cations (such as Zn^{2+}) dimers associate into hexamers (4), which is the configuration of insulin stored in the β-cells of the islets of Langerhans in the pancreas.

The islet cells make insulin by first synthesizing proinsulin, a peptide that contains the A and B insulin chains and an additional chain (C peptide), which is excised in most insulin molecules before secretion (5). Proinsulin itself may also play an important role in overall glucose metabolism.

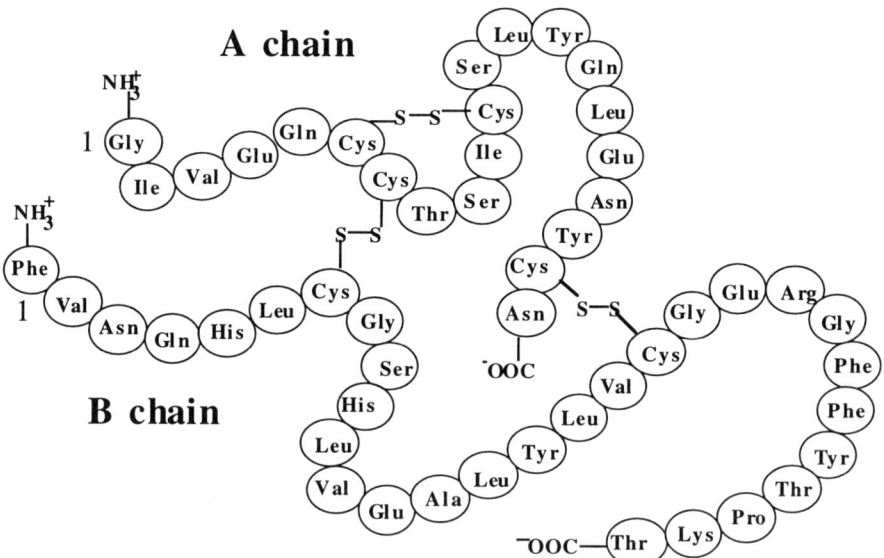

Figure 1. The primary structure of the human insulin monomer. The monomer consists of two chains, A and B, which contain 21 and 30 amino acids, respectively, and one intrachain and two interchain disulfide bonds.

Diabetes Mellitus. Diabetes mellitus is a complex disease thought to involve autoimmune, environmental, and perhaps viral components, in which the metabolisms of various substances are abnormal (6). Glucose is the most prominent of these, partially because of the ease with which it can be measured in the blood. In addition, abnormalities in glycogen, ketones, lipids, certain proteins, and even minerals, have been observed as signs in diabetic patients. If the situation is untreated, fat and muscle metabolism must replace carbohydrate metabolism resulting in high plasma levels of organic acids that can result in ketoacidosis and diabetic coma.

Diabetes affects millions of people worldwide. According to the American Diabetes Association (ADA), there are 14 million diabetics in the U. S. alone, only half of whom have been diagnosed (7). Diabetes is the leading cause of blindness and is one of the leading causes of death. Treatments for diabetes include administration of insulin, weight control, dietary control, and hyperglycemic drugs, such as sulfonylureas. According to the ADA study, about 1.5 million diabetics in the U. S. are treated via insulin therapy. All insulin-dependent (Type I) diabetes patients and some non-insulin-dependent (Type II) ones require an exogenous source of insulin. The Diabetes Control and Complications Trial Research Group (DCCT) has established that intensive insulin therapy, including frequent blood glucose monitoring and three or more insulin injections per day, significantly reduces the risk of developing long-term diabetic complications such as retinopathy, neuropathy, and nephropathy (8).

The Insulin Market. According to the Wilkerson Group (a biotechnology, drug, and medical consulting firm), the total U. S. market for drugs to treat diabetes was about $940 million in sales in 1990 (9). Insulin represents $465 million of this total with the remainder due to sale of oral anti-diabetic drugs. The growth for insulin sales is estimated at about 11% annually, projecting into $965 million in sales by 1997. In contrast, oral anti-diabetic drug sales are expected to decline at about 3% annually, resulting in sales of $390 million by 1997. Thus, insulin should remain the dominant drug in the treatment of diabetes.

Insulin is produced in a variety of formulations according to various U. S., British and European pharmacopoeias. A partial listing of some of the commercially available insulin preparations and their manufacturers is shown in Table I, adapted from ref. (10). These are classified into rapid-, intermediate-, rapid/intermediate- and long-acting. The two suppliers which dominate the U. S. market, Novo Nordisk (Denmark) and Eli Lilly (Indianapolis, IN), are represented.

Issues in Insulin Therapy. The typical therapy for about half of all insulin-dependent diabetics is a subcutaneous or intramuscular injection of concentrated insulin preparations. Because insulin has a limited lifetime *in vivo* (about 30 min in humans (11)), multiple injections are required. Both short-term and prolonged actions are desirable so some commercial preparations contain combinations of rapid-acting and long-acting formulations.

To efficaciously deliver insulin and achieve normoglycemia, the blood insulin level should mimic the normal one as closely as possible. However, it is difficult to mimic the normal, physiological profile by injection. Between meals, when the blood glucose level is 80-90 mg/dL, only a basal level of insulin is desired (a plasma level of about 10 μU/mL) (12). Following meals, the glucose concentration rises rapidly, and a much higher level of insulin (in the range of 100 μU/mL) is required (12). Thus, an optimal insulin formulation must have both rapid and prolonged actions. Following injection, there is a slow rise to a therapeutic level of insulin which may result in postprandial hyperglycemia, after which there is a slow decline in insulin concentration potentially leading to hypoglycemia between meals. Insulin absorption and dispersion rates vary not only because of differences in their formulations, but also due to differences in patient behavior, metabolism, and eating habits (13).

Table I. Examples of commercially available insulin preparations[a]

Class	Supplier	Name	Source	Onset (h)	Peak (h)	Duration (h)
Rapid-acting						
Regular insulin	Eli Lilly	Humulin R	H[b]	0.5	2-4	6-8
		Regular Iletin I	B,P	0.5	2-4	6-8
		Regular Iletin II	P	0.5	2-4	6-8
	Novo Nordisk	Novolin R	H	0.5	2.5-5	8
		Regular	P	0.5	2.5-5	8
		Velosulin BR (semisynthetic)	H	0.5	1-3	8
Insulin analog	Eli Lilly	Humalog	H	<0.25	0.5-1.5	3-5
Intermediate-acting						
Lente insulin (zinc spn)	Eli Lilly	Humulin L	H	1-3	6-12	18-24
		Iletin I Lente	B,P	1-3	6-12	18-24
		Iletin II Lente	P	1-3	6-12	18-24
	Novo Nordisk	Lente	P	2.5	7-15	22
		Novolin L	H	2.5	7-15	22
Isophane insulin	Eli Lilly	Humulin N	H	1-2	6-12	18-24
NPH (protamine spn)	Eli Lilly	Iletin I NPH	B,P	1-2	6-12	18-26
		Iletin II NPH	P	1-2	6-12	18-26
	Novo Nordisk	Novolin N	H	1.5	4-12	24
		NPH	P	1.5	4-12	24
Rapid/Intermediate-acting						
NPH/Regular	Eli Lilly	Humulin 50/50	H	<1	2-12	24
		Humulin 70/30	H	<1	2-12	24
	Novo Nordisk	Novolin 70/30	H	0.5	2-12	24
Long-acting						
Ultralente (zinc spn)	Eli Lilly	Humulin U	H	4-6	8-20	24-28

[a]Data for Humalog obtained from product insert obtained from Eli Lilly (Indianapolis, IN). All other data obtained from ref. (*10*).

[b]H=(recombinant) human; B=bovine; P=porcine; B,P=70% bovine and 30% porcine; spn=suspension.

In addition, patient compliance is a problem in insulin injection therapy. Typically, diabetics must inject themselves at least twice a day. They consider such injections uncomfortable at best and often painful. This leads to a high non-compliance rate, in the range of 40-50% (*14*). The large number of injections required is a particular problem for diabetic children. Due to these considerations, insulin is a candidate for novel modes of delivery (discussed below).

The insufficient stability of insulin formulations presents additional problems. Like all protein pharmaceuticals, insulin possesses relatively large molecular weight, contains multiple functional groups that are susceptible to deleterious processes, and has a three-dimensional structure that is critical for biological activity. Under certain conditions, insulin in solution is subject to non-covalent aggregation (*15*), e.g., 'frosting' of refrigerated commercial formulations (*16*). This aggregation poses major problems in alternative modes of insulin delivery. Implications of these instability pathways are discussed in more detail below.

HISTORICAL DEVELOPMENT OF INSULIN AND INSULIN THERAPY

Major events in the pharmaceutical development of insulin are shown in Figure 2. Some of the milestones reflect technical advances in protein biochemistry, while others represent the successful application of new methodologies of protein production, purification, and administration. Since the initial discovery of insulin, pharmaceutical researchers have competed to produce more efficacious insulin formulations and modes of delivery. Bringing these new products to market and clinical application has been an ongoing goal for many pharmaceutical companies.

The clinical development of insulin was also paralleled by advances in federal regulation of pharmaceuticals. As insulin gained more widespread use in the U. S. and worldwide, a need was perceived for increased federal regulation of insulin and other drugs. As a result, production of insulin preparations now falls under strict regulations. These regulatory conditions have fostered a market in which only a few large pharmaceutical companies are involved.

Discovery and Early Development. Insulin was first discovered in pancreatic extracts at the University of Toronto by Banting and Best (*17*). Dispute over claims of credit in the discovery of insulin led to a 1923 Nobel Prize being shared by Banting and Macleod, and furthermore, the prize money was shared with two other researchers, namely Best and Collip. An interesting review of the story of insulin discovery is presented by Bliss (*18*). Initially, the only formulation available was a crude solution of insulin derived from porcine, and later, bovine pancreas. This solution was acidic in order to maintain insulin solubility and inhibit the enzymatic action of numerous contaminating pancreatic enzymes. The earliest clinical trials that were conducted in the 1920's demonstrated that insulin was ineffective when delivered orally, rectally or intranasally. Thus, subcutaneous injection became the standard route of administration for insulin, and other protein pharmaceuticals.

Abel performed the first crystallization of insulin in 1926 (*19*), and Scott (*20*) successfully obtained insulin crystals in the presence of zinc. Following these advances, crystalline insulin suspensions were formulated for clinical use. These formulations were more pure than the previously used crude acid insulin solutions and had more reliable therapeutic action.

Following these early advances, researchers endeavored to prolong the activity of insulin. The first formulation with prolonged action put into therapeutic use was protamine zinc insulin (*21*). Protamine zinc insulin (PZI) was the first stable insulin crystal suspension. This formulation is prepared by precipitating insulin in the presence of zinc and another protein, protamine. Depending on the specific mode of preparation, PZI can be either intermediate-acting or long-acting.

1920 — Insulin is discovered

Crystallization of insulin
Insulin is a polypeptide

1930 — Crystallization of Zn-insulin
Protamine insulin
Protamine Zn-insulin

1940 — Federal Food Drug and Cosmetic Act
Stricter federal regulation of insulin production

Iso-insulin developed
Isophane insulin (NPH)

1950 — Re-crystallization reduces immunogenicity
Lente insulin
Insulin becomes the first protein sequenced
4-Zn insulin crystallization
Biphasic insulin (Rapitard)

1960 — Neutral insulin solution
Kefauver-Harris Drug Amendments

Discovery of proinsulin
1970 — First 3-dimensional crystal structure of insulin

Monocomponent insulin
Development of insulin infusion systems

1980 — Total chemical synthesis of insulin
Eli Lilly produces recombinant human insulin in *E. coli*
Semisynthetic insulin developed at Novo
Early attempts at controlled release of insulin
Novo produces a recombinant human insulin in yeast
Novo Nordisk develops a genetically engineered monomeric insulin
1990 — NMR solution structure of insulin
Eli Lilly develops insulin lispro, engineered for rapid action
On-going development of non-parenteral delivery routes

Figure 2. Overview of the pharmaceutical development of insulin.

Development of the Insulin Pharmaceutical Industry. The commercial importance of insulin was immediately recognized by the pharmaceutical industry. Thus, it is not surprising that several pharmaceutical companies became interested in developing new insulin therapies. Two major pharmaceutical companies in the early development of insulin were Danish firms, Novo and Nordisk.

In 1946 (22), researchers in Hagedorn's laboratory at Nordisk Insulinlaboratorium developed a stable PZI modification dubbed NPH (for Neutral Protamine Hagedorn). This formulation is produced by precipitating insulin and protamine in isophane (i.e., nearly equal) proportions; the preparation is also known as isophane insulin (23).

Hallas-Møller and co-workers at Novo laboratories developed the Lente insulins during the 1950's (24,25). These preparations do not employ protein precipitants. Instead, prolonged action was obtained by suspension of insulin crystals at neutral pH in the absence of any zinc-binding ions (such as phosphate or citrate) which were present in previous preparations. Increasing the amount of zinc in solution first forms an amorphous insulin precipitate (Semilente) and then crystalline insulin (Ultralente). Semilente exhibits rapid action and Ultralente exhibits more prolonged action.

Regulatory Issues Regarding Insulin Development. The initial clinical trials of regular (acid) insulin solutions had occurred before the formation of the Food & Drug Administration (FDA) in 1931. At that time, there were few federal regulations regarding the pharmaceutical industry in the U. S. (see ref. (26)). The Import Drug Act (27) ensured that all imported drugs were properly labeled and examined for their medical usefulness. The Pure Food and Drug Act (28) was passed to safeguard against mislabeling and adulteration of food and drugs produced within the U. S. Insulin, as a biological drug, was also regulated under the Biologics Act (29) which provided for all biological drugs to be licensed and produced only by licensed manufacturers.

Early in the clinical development of insulin (1938), the Federal Food, Drug and Cosmetic Act (30) was passed, establishing the concept of "new drugs" which had to be proven safe before they could be sold. Pharmaceutical companies were required to file an investigational new drug (IND) report. In addition, before any drug could be marketed a new drug application (NDA) had to submitted to and approved by the FDA. These regulations later became more strict under the Kefauver-Harris Drug Amendments (31) which provided for more severe safety requirements and for a satisfactory demonstration to the FDA of drug efficacy.

As use of insulin became widespread in the U. S. and worldwide, there was a growing concern over its regulation. As a result, in 1941 Congress added Section 506 to the 1938 Federal Food Drug and Cosmetic Act (32). This law required that the federal government test and certify each and every batch of insulin before it reaches the U. S. market (other drugs were not, and still are not covered by such rigid requirements). This was followed later by similar amendments covering penicillin (33) and other antibiotics (34-36). The loosening of such requirements is part of recently proposed reforms in regulations regarding biologicals (37).

As a result of these developments it became more difficult for small pharmaceutical firms to develop new insulin formulations and other anti-diabetic drugs (see ref. (38)). Due to the new regulations and the increased costs of preclinical testing, there were few "me-too" insulin formulations developed by small drug firms. As a result, several large pharmaceutical companies comprised an oligopoly on insulin products.

Development of Highly Purified Insulins. Clinical potency of any drug is dependent upon its purity, and this has proven to be especially true for pharmaceutical proteins like insulin. The first insulin preparations for clinical use were very crude by modern standards. Impurities present in insulin formulations lead to less specific potency, greater instability and increased immunogenicity. It was observed by Jorpes (39) that patients who suffered allergic reactions to insulin had better tolerance when their insulin had been re-crystallized up to seven times. With the advent of new analytical techniques such as high performance gel filtration chromatography and disc electrophoresis, it became clear that many commercially available insulin preparations had non-insulin material. It was later shown by Schlichtkrull and co-workers (40) that these non-insulin impurities were causing the immunogenic reaction.

Pharmaceutical companies have taken advantage of advances in protein purification technology to produce highly purified insulins (see Brange (41)). The characterization of these preparations is based on several analytic methods as reviewed by Jørgensen et al. (42). One strategy to further purify insulin employs multiple chromatographic steps such as gel chromatography and ion-exchange chromatography. Some notable examples of highly purified insulin preparations that have been developed include monocomponent (MC) insulin (produced at Novo), single component (SC) insulin (Eli Lilly), and porcine RI insulin (Nordisk Insulinlaboratorium). Some drug companies have also produced insulin purified by a single chromatographic step designed to remove the non-insulin components. These products include SP-insulin (Eli Lilly), CS- and CR-insulin (Hoechst), and mono-pic insulin (Organon). The clinical usefulness of these highly purified insulins has been discussed elsewhere in detail (43). Even though most insulin in use today is purified by chromatographic techniques, some pharmaceutical companies still produce insulin purified only via crystallization.

Biphasic Insulin and Surfen Insulins. A need for an insulin preparation with both rapid-acting and intermediate-acting components led to the introduction of Rapitard insulin (44,45). Rapitard is a mixture of crystalline bovine insulin (the intermediate-acting portion) and dissolved porcine insulin (the rapid-acting component). This formulation is based on the differences of bovine and porcine insulin solubilities as a function of pH and zinc concentration. Formulations containing both dissolved and crystalline insulin are known in general as biphasic preparations. Examples of biphasic insulin formulations that have been developed include Initard and Mixtard (Nordisk), both composed of a mixture of regular (acid) dissolved insulin and NPH insulin.

Another approach to obtaining insulin formulations with both rapid and intermediate or prolonged actions is the combination of regular insulin with insulin that has been complexed with the urea derivative surfen (46). The surfen insulin precipitates as amorphous particles after injection and has longer duration of action. Examples of biphasic surfen insulins include Komb-H Insulin and Depot-H Insulin (Hoechst). Surfen insulins have been widely used in Germany, but increasing concern over incidence of allergic reactions to surfen has retarded its application in the U. S. and elsewhere (47).

Investigations of Insulin Structure. Insulin has been the focus of intense study since its discovery. As early on as 1928, it was realized by Wintersteiner et al. (48) that insulin was a protein. Later, pioneering work by Sanger and others on the elucidation of the chemical nature of proteins resulted in the determination of the primary structure of porcine insulin (49). This discovery represented another milestone in the history of insulin research: insulin had become the first protein to be sequenced.

In addition, the three-dimensional crystal structure of insulin has been extensively studied and has been determined for various crystals from porcine pancreas such as 2-zinc (*50-52*) and 4-zinc (*53,54*) porcine insulins. The three-dimensional structure of porcine zinc-free insulin has also been determined (*55*). Insulin crystals used for injection therapy can be amorphous (e.g., Semilente), 2-zinc (e.g., Rapitard) or 4-zinc (e.g., Ultralente) in structure. The three-dimensional crystal structure of human insulin has also been elucidated and is nearly identical to the 2-zinc (*56*) and the 4-zinc (*57*) porcine forms.

More recently, the structure of insulin in solution has been determined using two-dimensional NMR. Insulin solution structure is nearly identical to the crystal structure (*58,59*). This result is significant in that it validates the application of crystalline structural data for developing new aqueous insulin formulations for pharmaceutical use.

The structures of insulin from various other species have also been determined, and the knowledge gained in this work may be useful in designing new insulins via protein engineering. For example, it has been shown that the insulin monomers from some species do not associate even in the presence of zinc (e.g., ref. (*60*)) and study of these insulins may be useful in obtaining monomeric insulins that have rapid action (see below).

Another related area concerns the insulin receptor. In 1985, the insulin receptor was cloned and sequenced (*61,62*), and more recently the crystal structure of the tyrosine kinase domain of the human insulin receptor was determined (*63*). It is hoped that such investigation will lead to further understanding of the molecular basis of insulin action and may be useful in genetically engineering more potent insulins. Although it was long believed that the effects of insulin are mediated by a unique receptor, there is considerable evidence that insulin receptors in the brain, liver, adipocytes, and lymphocytes are heterogeneous in structure and function, commensurate with insulin's multifaceted action (*64*).

Insulin Derivatives. Researchers have also investigated chemically modified versions of the insulin molecule. These studies have helped to elucidate the spatial organization of the insulin molecule and the role of various functional groups in stabilizing its structure (*65*) and have revealed the biologically active site on the insulin molecule (*66*).

Of direct relevance to the pharmaceutical industry were attempts to produce longer-acting insulins, and less immunogenic preparations of insulin. Clinical development of an insulin derivative was first explored when Hallas-Møller (*67*) chemically linked insulin with phenylisocyanate (Iso insulin) which was soluble at physiological pH and demonstrated prolonged action. Other efforts targeted residues implicated in immunogenic studies, such as the carboxy-terminal amino acid of the B chain (*68*).

Pharmaceutical Development of Proinsulin and C-Peptide. Some 45 years after the discovery of insulin, details were uncovered regarding its biosynthesis. The biosynthesis of insulin from proinsulin precursor was first described by Steiner and Oyer (*5,69*). In the endoplasmic reticulum of β-cells of the islets of Langerhans, a single-chain polypeptide, proinsulin, is transcribed and stored in the Golgi apparatus (*70,71*). This proinsulin peptide is then cleaved by a membrane-bound protease to yield insulin and connecting peptide (C-peptide).

Interest in the pharmaceutical industry has resulted in recombinant human proinsulin for clinical study (*72*). Eli Lilly was among the first to successfully produce recombinant human proinsulin (*9*). Early in the clinical trials, it was revealed that in one study there were more heart attacks in a diabetic group receiving the proinsulin than in a control group (*73*). Even though the correlation with injected proinsulin remains uncertain, Eli Lilly decided not to pursue the development of this product.

In addition, the C-peptide has some potential for clinical use. It was long-believed that C-peptide did not exert any biological effect based on a small number of trials in humans (74,75). However, more recent studies suggest that C-peptide administration to patients with insulin-dependent diabetes on a short term basis (3 hours) may augment whole-body glucose utilization, ameliorate autonomic neuropathy, and reduce glomular hyperfiltration (76).

Production of Human Insulin. Perhaps the most significant advance in diabetes therapy since the initial discovery of insulin has been the production of recombinant human insulin. The use of human insulin over other forms offers significant advantages; for a review with respect to human insulin's biological activity, pharmacokinetics, and therapeutic use, see Brogden and Heel (77).

Starting in the 1970's, there was growing interest in producing human insulin for therapeutic use. Because of the scarcity of human pancreas from cadavers, human insulin must be synthesized to produce the large amounts desired clinically. Progress in protein synthesis from amino acids led to the total synthesis of human insulin (78,79). However, this approach is not industrially useful due to high production costs and difficulty in producing correctly folded, disulfide bonded insulin monomers.

It is also possible to chemically alter porcine insulin to obtain the human form (80). Human insulin differs only in the carboxy terminal residue (amino acid 30 of the B chain) where Ala^{B30} in porcine insulin is replaced by Thr^{B30} in the human form. The development of this 'semisynthetic' human insulin (by Novo) allowed for the production of clinically sufficient quantities. This strategy has been useful clinically, but it still does not resolve the issues of rapidly growing demand and a limited supply of porcine insulin.

Recombinant DNA technology has revolutionized the clinical development of human insulin. Such advances have ensured an almost unlimited supply of highly purified human insulin. Two major pharmaceutical companies were initially involved in the race to produce recombinant insulin, first Eli Lilly in collaboration with Genentech, and later, Novo. An interesting account of this historic pursuit in biotechnology is presented by Hall (81).

The approach employed at Eli Lilly was to produce the A and B insulin chains by cloning them separately in *E. coli* (82). Through a series of elaborate post-fermentation chemical modifications, the chains are then recombined and the proper disulfide bonds are formed to yield the native insulin molecule. Another method involving cloning of human proinsulin (83) has a much simpler recombination process. Using this technique human insulin of 97% purity can be obtained. Clinical trials by Eli Lilly in 1981 marked the first pharmaceutical use of a recombinant human protein.

Production of recombinant human insulin in *Saccharomyces cerevisiae* has also been reported (84,85). Both A and B chains are cloned together with only a few additional amino acids between them. The result is an insulin-like molecule that is exported extracellularly, and folds correctly with the correct disulfides. The additional residues are then removed. Following purification, this methodology can result in an insulin purity of up to 99.9%

Genetic Engineering of Insulin. Conceiving new and effective rapid-acting and prolonged-acting insulin formulations still remains a challenge today. Previously, most research efforts to improve insulin were rather unsystematic. Current studies take a more rational approach, building on the wealth of information on the insulin molecule and the tools of recombinant DNA technology.

Researchers at Novo Nordisk have identified several key amino acid residues in the insulin molecule that are important for association into hexamers (86). Their rational approach was to use the X-ray structure of the insulin hexamer to discern interactions between insulin monomer in dimers and hexamers and to predict the alterations necessary to disrupt these interactions. The goal, obtaining monomeric insulin, will allow for faster onset of action, administration of insulin with a meal without losing glycemic control, and the potential to minimize late hypoglycemia. The residues involved are mainly B-chain residues, some of which are also in the putative receptor-binding region. These residues include Ser^{B9} and Thr^{B27}. Replacement of these residues with aspartic acid and glutamic acid resulted in less associated insulin that had a more rapid action (86). Using this strategy, Novo Nordisk has introduced Actrapid HM insulin in the European market and has started clinical testing in the U. S. Another variant, wherein Tyr^{B16} has been replaced by His^{B16}, is monomeric at millimolar concentrations in aqueous solution at low pH, which allowed for its structure determination by NMR (87). This structure showed only subtle differences from that of the X-ray crystal structure of the native hexamer. Novo Nordisk is also working on producing longer-acting insulins. In one project under development, the residue Thr^{B27} has been replaced with arginine and the carboxyl terminus at Thr^{B30} derivatized with an amide. This results in an insulin that is soluble in acidic solution but will crystallize at the higher pH of plasma. These genetically engineered insulin molecules, and more than 30 others, have been reviewed by Brange et al. (88).

Researchers at Eli Lilly have taken a similar approach to develop a rapid-acting insulin analogue (89,90). Their cue came from insulin-like growth factor (IGF-1), which has considerable (50%) structural homology with insulin (especially in the C-terminus of the B chain) yet does not self-associate. In particular, IGF-1 has Lys^{B28}-Pro^{B29}, whereas these residues are reversed in the native insulin sequence. It was found that reversing these residues in the native insulin sequence resulted in a monomeric form with more rapid action than (unmodified) neutral regular human insulin (91,92), improving the control of postprandial hyperglycemia in insulin-dependent diabetic patients (93). Following clinical testing, in 1996 the FDA granted its approval (94) for the Lily insulin analogue, also known as insulin lispro. It is also being developed as a crystalline protamine suspension, NPL (neutral protamine lispro) exhibiting an intermediate-acting physiological profile analogous to NPH (95), and in a biphasic mixture (of soluble insulin lispro and NPL) with both rapid- and intermediate-actions (96).

DEVELOPMENTS IN INSULIN DELIVERY

Ongoing pharmaceutical investigation of insulin is based not only on improving its quality for injection therapy but also on developing new administration routes. These include nonparenteral routes (e.g., pulmonary), insulin infusion systems and controlled-release from polymers. Such studies demonstrate the delivery and stability problems generally encountered with protein-based pharmaceuticals.

Parenteral Delivery (Syringe). The need for chronic administration and a continuously variable dosing regimen make the formulation and delivery of insulin a particularly challenging problem. The classical method of achieving glycemic control in patients with insulin-dependent diabetes is via subcutaneous or intramuscular injections of concentrated insulin preparations (41). The preferred regimen is based on at least two daily injections (before breakfast and at bedtime). Mixtures of rapid-, intermediate- and long-acting (slowly dissolving) insulins are used to counteract the increased blood sugar content at meal times and to maintain a basal insulin concentration between meals. This relatively simple and successful approach belies the many difficulties associated with the treatment of this autoimmune disease.

The DCCT Research Group has established that intensive insulin therapy—including frequent blood glucose level monitoring and three or more insulin injections per day—significantly reduces the risk of developing serious long-term diabetic complications such as retinopathy, neuropathy, and nephropathy (8). Even so, this intensive subcutaneous treatment is not ideal. Sporadic glucose sensing does not provide diabetic patients sufficient information to meet actual, changing insulin demands. Furthermore, injecting insulin into the peripheral circulation is non-physiological. Normally, the pancreas secretes insulin into the portal circulation to inhibit glucose production by the liver (where 50% is degraded by first-pass metabolism), and the remainder proceeds through the peripheral circulation to increase cellular glucose utilization. Injected insulin, on the other hand, mainly affects the peripheral glucose uptake and not regulation. Also, subcutaneous treatment is imprecise in both estimating and delivering the insulin dosage. Differences in patient behavior, metabolism, insulin antibody levels, eating habits and physical exertion, as well as injection location and technique (including angle and depth, and handling of insulin solutions) all contribute to variability in the absorption rate (*13,88,97*). For example, it has been reported that Type I diabetic children frequently lose insulin at the site of injection, which can lead to a dangerous variability in blood glucose (*98*). Finally, occasional hypoglycemic episodes and general hyperinsulinemia, known to be associated with weight gain and an increased risk for atherosclerosis, are inevitable with subcutaneous treatment (*99,100*).

Pen and Jet Injectors. In addition to the problems discussed above, injection therapy discriminates against those either unable or unwilling to self-administer the injections routinely. Increasing patient compliance would improve diabetes treatment. One approach is to make insulin injections less painful and more convenient. To eliminate the anxiety associated with needles, needleless jet injectors can be used, which force insulin through a fine nozzle at a high pressure. For those who prefer the conventional needle administration, a pen injector can offer increased convenience.

Insulin delivered by jet injector exhibits different absorption kinetics than that administered by needle, providing a faster and shorter hypoglycemic effect due to dispersion over a larger area within subcutaneous tissue (*101*). This necessitates a multiple injection regime, promoting improved long-term health at the expense of more attention from the patient. Relatively faster kinetics of insulin action also allow preprandial injections to be closer to meals (subcutaneous injections are often administered 30-45 min before eating to match the elevated glucose levels at meal times). Patient surveys have generally supported the jet injector (*101,102*), referring to a decreased (though not minimal) level of pain associated with injection. One might expect that the high pressure and shear force involved in jet injection denatures insulin, thereby increasing the antibody response. On a practical level, one might also expect that jet injection requires extensive education to ensure accurate delivery to avoid leakage, hematomas, and bleeding. However, one study reported a decreased antibody response and a shorter time to properly train patients to administer insulin by jet injector compared to syringe (*103*). Jet injectors have also been determined to be mechanically reliable and extremely dose-accurate (*104,105*), as well as unlikely to be a potential source of infection (*106*). The unit is not universally accepted, though, perhaps because of its cost (although medical insurance may help cover the expense), its low publicity, or simply because some patients prefer the method of conventional injections.

Multi-dose pen injectors provide the convenience of insulin treatment without changing the kinetics of conventional subcutaneous delivery. This option offers good metabolic control, encourages a multiple injection regime, and provides a simple, reliable dosage form. The system is associated with an insulin loss of about 1% (*105*), an improvement over the 5-19% associated with a conventional syringe (*107*), provided that the patient follows proper techniques for injecting the full dose and for preventing the incorporation of air bubbles into the cartridge (*108*). Pre-filled cartridges do not allow for the mixing of different types of insulin. At a total cost similar to that of syringes and insulin vials, the pen injector/cartridge system has become increasingly popular.

In terms of effective treatment, however, neither jet nor pen injector provides significant clinical improvement over the conventional multiple injection regime by syringe. Although the important issue of patient compliance is addressed, injectors are still subject to the same problems correlated with non-physiological subcutaneous treatment.

Continuous Subcutaneous Insulin Infusion (CSII). External insulin infusion pumps sidestep the need for multiple injections altogether, and provide a benefit in that the continuous flow of small amounts of the hormone more closely mimics normal release. Unfortunately, the improvements provided by pump are not very significant in comparison to multiple subcutaneous injections. Moreover, proper use of the external insulin pump requires much education and attention, so it is neither convenient nor appropriate for all diabetics.

Infusion systems have been investigated which regulate insulin delivery based on measured glucose concentration (*109,110*) and those which deliver insulin according to an inputted regime (*111,112*). Although such devices are available on the market, substantial modifications are required for wider acceptance. The major stumbling block in this technology is insulin stability in neutral agitated solutions. The environment inside an infusion device often leads to insulin aggregation which results in potency loss and clogging of the device's tubing and catheters. Mechanistic investigation of insulin aggregation may eventually solve this problem and lead to successful application of infusion systems on a larger scale.

The advantages of CSII appear to weigh favorably against the disadvantages given an educated, diligent patient and a reliable infusion device. After thorough communication between the patient and an experienced clinician has been established (*113*), the pump can remove the variables involved with conventional injections, providing an accurate, predictable insulin absorption and stable glycemic control (*114*). External pumps currently on the market are lightweight (about 3.5 ounces), durable, easy to use, have soft catheters to replace needles, as well as tubing designed to minimize insulin aggregation and subsequent device obstruction. A programmable feature on the pumps also allows the patient to meet insulin needs throughout the day, thereby satisfying circadian insulin demands such as the pre-dawn phenomenon of hypoglycemia (*115*). Expectedly, the use of an insulin pump results in less glycemic fluctuation since insulin peaks do not fade away over time as with sporadic injections.

Initiation of intense CSII treatment is associated with short-term worsening of retinopathy (*8*), and an abrupt initiation or discontinuation of the pump has been known to cause ketoacidosis (*116*). However, it has been shown that patients who are carefully selected and educated show at least the same frequency of ketoacidosis (*117*), if not lower (*118*), compared to those using conventional injection therapy. Examining other important markers of diabetes, it has been shown that CSII does not significantly improve glycosylated hemoglobin values, reduce hypoglycemia, or prevent weight gain associated with multiple subcutaneous injections. However, fasting lipid levels (*119,120*), glucagon profiles (*121*), and motor nerve condition velocities (*122*) were improved when CSII was used in place of subcutaneous insulin injection. The benefits and satisfaction gained from CSII depend on the diligence of the patient. At least four daily assessments of blood glucose levels are required, and precautions such as changing infusion sets, disinfecting the infusion site and maintaining the pump are crucial. As with the injectors, surveys show that CSII does not draw strong preferences from all patients, although some Type I and Type II diabetics strongly support it (*120,123*). The next major advance, which would likely result in more widespread use of CSII, would be the incorporation of a glucose sensor directly in the pump.

Implantable Intraperitoneal Pumps. Implantable intraperitoneal pumps (IIP) best address the need for a more physiological delivery of insulin, while also decreasing the time and variability of absorption involved in subcutaneous administration. For a review of implantable insulin pumps and their metabolic control, see Hepp (*124*).

Compared to subcutaneous delivery, insulin introduced into the peritoneal cavity is absorbed more rapidly (*125*), results in lower peripheral insulin levels (*126*), reduces the amount of high-density lipoproteins (*127*), and normalizes lactate levels and plasma insulin kinetics (*127*). Improved regulation of hepatic metabolism and increased hepatic extraction lead to lower peripheral insulin levels, thus reducing the risk of hyperinsulinemia (*128*). Consistent glycemic control results in fewer episodes of hypoglycemia and ketoacidosis.

The feasibility of successfully implanting a pump under subcutaneous fat of the abdomen has been demonstrated in several clinical trials (*129-131*). Early studies were plagued by a variety of clinical and technical problems (*129*), but improvements in pump design and the insulin formulation significantly reduced the number of complications (*130*). The health benefits of implantable pumps have been demonstrated in a recent study: a six-fold decrease in the number of incidents of severe hypoglycemia was reported when IIP was compared to a multiple injection regimen (*131*). These benefits were confirmed in long-term follow-up studies which also reported lower glycosylated hemoglobin levels and fewer incidences of ketoacidosis, suggesting that IIP reduces glycemic fluctuations (*132*).

In an interview with patients who had implanted pumps (*133*), many cited better well-being and physical and mental health, and none regretted their decision for implantation (the major complaint was the inconvenience of pump refilling). However, pump reliability and safety issues need to be more thoroughly considered before widespread, approved use. Complications requiring removal and other surgical interventions were encountered in a recent feasibility study (*134*). Problems included skin atrophy and catheter obstructions, the latter a particularly significant problem when polyethylene catheters were used (*135*). Most obstructions were due to fibrin clots or tissue encapsulation, with a smaller percentage possibly due to insulin aggregation. The only apparent risk factor for complications at the implantation site was heavy physical activity. Patients with a previous experience of peritoneal insulin infusion from portable pumps and a duration of diabetes of 21 years or greater were more likely to develop complications with IIP. A final question regarding the long-term safety of implantable pumps arises from a noted increase in insulin antibodies in some patients (*136,137*). Further study is required to determine the consequences of elevated levels of insulin antibodies.

Controlled-Release Devices. Polymeric devices have also been investigated as alternative insulin delivery systems. Aside from the gain in patient compliance, this approach would achieve sustained- or controlled-release of insulin. However, many issues need to be considered in the development of polymeric release of insulin, including biocompatibility and the toxicity of degradation products (for bioerodable matrices) (*138*). Furthermore, there is the potential for insulin degradation in the polymer, either during the device processing/sterilization, or during *in vivo* release.

In either injectable or implantable polymeric devices, unregulated release could supply a basal dose of insulin, whereas externally-regulated release could potentially meet pulsatile demands. Even more attractive is the prospect of self-regulated, glucose-dependent insulin release. Although implantation of such systems may involve surgical procedures, polymers represent a potentially more biocompatible, comfortable, energy independent, and space efficient option than implantable pumps. The risks of unexpected pump failure and leakage could be eliminated, as could the need for surgical removal of the device in the case of a biodegradable system.

Unregulated Release. Methods devised for the continuous release of insulin include injectable biodegradable polyanhydrides (*139,140*), implantable or injectable biodegradable poly(lactic acid) pellets (*141,142*), an implantable ethylene-vinyl acetate matrix (*143-145*), zinc calcium phosphorous oxide ceramics (*146*), hydroxyethyl methacrylate (*147*), insulin-albumin microbeads (*148*) and gelatin films (*149,150*).

Biodegradable polymers may erode by hydrolysis, solubilization, or enzyme digestion into biocompatible products. *In vivo* studies have shown that biodegradable systems can be used to deliver efficacious doses of insulin over a period of several days without causing inflammation (*140,141*). Furthermore, it was shown that implanted insulin-albumin microbeads achieved glycemic control in animal models for about 2 weeks, with complete biodegradation occurring over 4 to 8 weeks (*148*). Some non-degradable systems were also tested in animal models, where they maintained a constant level of insulin for up to 20 (*146*) or even 40 (*145*) days. One drawback with non-degradable systems, however, is the need for their removal once release rates are no longer constant.

Externally Regulated. Several approaches are being investigated for the external regulation of insulin from polymer matrices. For example, magnets dispersed in alginate spheres were incorporated into an ethylene-vinyl acetate copolymer, which, in the presence of an oscillating magnetic field, increased the rate of diffusion through the pores of the matrix (*151*). This system decreased glucose levels in diabetic rats by 30%, with no significant inflammatory response. In another study where the magnetized alginate spheres were dispersed in poly-L-lysine (*152*), it was shown that the polymer swelled in the presence of the oscillating magnetic field, and insulin diffused out upon the removal of the field. This effect was reproduced for up to 50 cycles *in vitro*, demonstrating the feasibility of the approach for longer-term insulin delivery. Other approaches to regulate insulin release from polymers include ultrasound (*153*) and compression (*154*). The feasibility of these strategies has been demonstrated in diabetic animal models (*153,155*).

One limitation of external insulin regulation from polymers is the lag in response (> 1 hour), which needs to be reduced. In addition, the toxicity of the proposed systems must be addressed. These issues represent significant challenges for future work in this area.

Self-Regulated Systems. Glucose-dependent polymeric release of insulin could mimic the normal production and secretion of insulin from pancreatic β-cells. One promising approach is based on the activity of the enzyme glucose oxidase, which converts glucose to gluconic acid. Incorporation of glucose oxidase into an insulin-containing polymer can facilitate the regulation of insulin release by controlling the device's pH. Increasing levels of gluconic acid produced at elevated glucose concentrations lower the pH of the polymer microenvironment, and studies have focused on translating this pH change into increasing the release of bound or entrapped insulin. This method takes advantage of the dependence of insulin solubility on pH to regulate its diffusion (*156*). Use of a modified insulin with a pI of 7.4 (instead of its normal value of 5.3), increased the system's performance much more significantly than changes in enzyme concentration, pore length, particle size, and protein loading (*157*). Others have investigated co-encapsulation of insulin and glucose oxidase in pH-sensitive liposomes consisting of β-palmitoyl-γ-oleoyl-L-α-phosphatidylethanolamine and oleic acid (*158*). *p*-Nitroacetanilide (a polymer which erodes in response to a change in pH (*159*)) and 2-hydroxyethyl methacrylate hydrogels containing immobilized glucose oxidase (*160*) have similarly shown adequate sensitivity to increased glucose concentrations. An alternative approach utilizes glucose oxidase to operate a chemical gate, which is activated following the reaction of glucose oxidase with glucose, facilitating insulin release (*161*). Glucose oxidase has also been used in a protein delivery device where a disulfide bond linking insulin to glucose oxidase is broken by electrons generated during the oxidation (*162*). Yet another self-regulating hydrogel insulin delivery system is being pursued by Peppas and co-workers (*163*). This hydrogel is comprised of poly(diethylaminoethyl methacrylate-ethylene glycol) containing functionalized glucose oxidase, into which insulin is loaded. *In vitro*, this system swells in the presence of glucose, releasing the insulin.

Self-regulating systems employing glucose oxidase have yet to be tested in animal. Long-term biocompatibility, as well as operational issues, remain to be solved. One major requirement is a constant supply of oxygen to the glucose oxidase, ensuring its activation. Furthermore, the sensitivity of these systems to the glucose concentration must be honed, in order that the blood's buffering capacity does not interfere with the release of insulin. Some progress has been made in developing implantable glucose oxidase-dependent systems to measure blood glucose levels, rather than to release insulin. For example, the feasibility of implanting an intravenous glucose sensor in dogs has been reported (*164*).

Another major area of research in self-regulating insulin delivery systems involves concanavalin A (con A). Con A is a protein which binds to glucose and other saccharides (*165*). Con A can also reversibly bind to glycosylated insulin. In this case, free glucose will compete with con A-bound glycosylated insulin, facilitating the latter's release. Several forms of glycosylated insulin with different con A binding properties have been tested (*166-169*) in an attempt to identify one whose glucose-dependent release most closely resembles the physiological response. Among several proposed glycosylated insulins tested by Kim and co-workers (*170,171*), succinylamidophenyl-α-D-glucopyranoside insulin (SAP-G insulin) was found to show the best combination of steady release and sensitivity to glucose concentration. SAP-G insulin did not elicit an antigenic response *in vivo* (*169*) and the modified protein retained its bioactivity while exhibiting increased stability (*172*). The performance of SAP-G insulin was improved by binding it to cross-linked con A microspheres, thereby decreasing the time to respond to the glucose concentration (*173*). An alternative method attaches con A to native insulin (*174*). In this system, dextran, which blocks membrane pores and prevents escape of insulin from an inner reservoir, is used to compete with glucose for the binding site on con A (*174*). Yet another competitive-binding system, which does not involve con A, depends on glucose binding to a phenylboronic acid moiety, which decreases the viscosity of a polymer gel, thus allowing entrapped insulin to diffuse out more easily (*175*). To develop glucose binding-dependent insulin release vehicles, further work is required to adjust insulin release to match physiological requirements, formulate a method to replenish the hormone, and assess the long-term safety of the overall implant.

Another interesting approach is the use of microencapsulated islets of Langerhans themselves as a self-regulating insulin delivery system. For example, Goosen and co-workers (*176*) have created a 'bioartificial pancreas' consisting of the islet cells within microcapsules of alginate and poly-L-lysine. Others have encapsulated the islet cells within myoblasts prior to implantation, resulting in increased biocompatibility and a prolonged life of the implant (*177*). In animal models, a single intraperitoneal transplant of insulin-containing microcapsules was successful in reversing the diabetic state for a prolonged period (*176*).

Aerosol Formulations (Nasal Delivery). Research in nasal delivery has proven relatively successful and promising with a variety of pharmaceuticals (for a review see ref. (*178*)). This simple delivery route is useful for several low-molecular-weight drugs, which absorb into the extensive microcirculation network underneath the nasal mucosa. Insulin can also be delivered this way. However, hyperinsulinemia remains a problem, poor membrane permeability requires the use of irritating enhancers, and dosage is removed rapidly due to mucociliary clearance. Furthermore, variability in delivery may result from colds, allergies and congestion.

Poor membrane permeability is a common problem for the delivery of large-molecular-weight drugs. This situation may be ameliorated by enhancers, which unfortunately tend to be toxic or irritating. Despite evidence of mucousal damage and mild to moderate swelling and vacuolation of epithelial cells found during histological studies of bile salts in rats (*179*), nasal enhancers have been tested in humans (*180-192*). In healthy adult volunteers, an aerosol dose of 0.5 units (U) insulin/kg plus enhancer typically led to a 50% drop in plasma glucose levels within 30-45 min.

Trials in human diabetics revealed several further complications with nasal delivery, including significant variability in plasma glucose levels following insulin administration. For example, in a small study with five Type I patients (who were given three doses of intranasal insulin), intra- and inter-patient variabilities in the relative decrease of glucose levels were 17% and 45%, respectively (*188*). Some of this inconsistency is inevitable, due to variable properties of the nasal mucosa.

One other concern is whether nasal delivery has the ability to mimic a physiological response to a meal. Studies with diabetics show a greater lag time for the peak hypoglycemic effect compared to healthy volunteers (*182*). This effect is also seen with intravenous treatment; the delayed glucose effect may be due to abnormal glucose counter-regulation in diabetics (*193*). Even though the lag time cannot be easily controlled, nasal absorption may still be an improvement over preprandial subcutaneous injection.

The irritating effects of enhancers present additional obstacles to successful implementation of nasal delivery, and a satisfactory compromise between enhancer efficiency and safety is critical. Various investigators report that bile salts of oxycholate and cholate both caused a burning sensation (*194*), and spraying a 1% Laureth-9 solution nasally caused irritation in half of the subjects tested (*182*). In contrast, a lecithin-based system caused mild, transient irritation in only 4 out of 68 patients (*192*). Other enhancers studied in animal models include cyclodextrins (*195-197*), bioadhesive-acting degradable starch microspheres (*198*), medium-chain fatty acid salts (*199*), sodium taurodihydrofusidate (*200*), and ammonium glycyrrhizinic/glycyrrhetinic acid in polymer spray formulations (*201*). However, these animal results cannot reliably predict enhancer efficacy in humans, due to significant differences in inter-species absorption. For example, a certain cyclodextrin mixed with insulin in liquid form caused complete nasal insulin absorption in the rat (*195*), but led to no absorption in humans (*196*).

One method to partially skirt the irritation issue of enhancers may be to dose more frequently and use a lower concentration of enhancer. This approach may be promising, considering that a 120 U dose resulted in a more pronounced hypoglycemic effect when delivered as two doses of 60 U at 20 min apart (*192*). Even so, nasal delivery of insulin still has drawbacks. Colds and allergies may adversely influence absorption through the nasal mucosa, further contributing to inter- or intra-patient membrane variability. However, it has been shown that the volume of dose and rate of inhalation do not contribute to variability (*202*). In addition, the risk of hyperinsulinemia remains since the drug is presented to the peripheral circulation, creating an non-physiologically high ratio of peripheral-to-portal insulin. Compared to subcutaneous treatment, the advantages of nasal delivery are significant, provided that reliability is maintained. Intranasal insulin could present an alternative means of providing a convenient and effective dose to complement injections of a basal insulin level, but it is clear that many issues still need to be addressed.

Pulmonary Delivery. An attractive alternative to nasal insulin delivery is presented by pulmonary absorption. A large surface area, extensive vasculature, a relatively thin, permeable membrane, and low extracellular enzymatic activity favor rapid and significant absorption. Variability in absorption still exists, but is not subject to the many different membranes and proteolytic enzymes which contribute to variability in other delivery routes, such as oral (see below). Pulmonarily administered insulin has been shown to retain most of its bioactivity in rabbit lung homogenates (203), so that metabolism in the lung may not be a limitation in delivery at all. Studies comparing aerosol and instillate intra-tracheal administration have demonstrated a hypoglycemic effect for both cases (204). One significant difference, however, is the apparent requirement for an absorption enhancer or protease inhibitor for the instillate formulation (205). For instance, in rabbits, the aerosol form of insulin was about ten times more bioavailable than the instillate form without an enhancer (204), although an instilled liposomal formulation provided a more significant hypoglycemic effect (206). When oral inhalation of aerosolized insulin was investigated in type II human patients, plasma glucose levels dropped 45% in 40 minutes if insulin deposition was maximized by regulating the aerosol flowrate and particle size (207). Even more encouraging were the high patient acceptability and the modest dose of insulin (~ 1 U/kg) required.

The development of pulmonary delivery of insulin has progressed to the clinical trial stage. Pfizer (Groton, CT), in partnership with Inhale Therapeutics (Palo Alto, CA), is currently conducting Phase II trials of a powder-based pulmonary insulin product (208). The Phase I results in 24 healthy volunteers compared single doses of aerosolized insulin to standard subcutaneous administration, and showed that the hormone was absorbed systemically and successfully lowered blood glucose levels (208).

Transdermal Delivery. Transdermal drug delivery enables easy adjustment of an administered dose, does not introduce the risk of infection, and potentially allows for sustained release of drug. Yet, it has been one of the least researched non-invasive methods of delivery, due to the extremely low permeability of the skin. Bromberg and Klibanov (209) demonstrated that this may be ameliorated by use of certain organic solvents, e.g., ethanol, which greatly facilitate insulin's transport across biomimetic membranes. This effect may be further enhanced by complexation of the protein with hydrophobic ion-pairing detergents (209). Recently, another approach towards transdermal delivery has been demonstrated using lipid vesicles (210). This technology is being developed by IDEA GmbH (Germany) and has shown promise for the transdermal delivery of insulin.

To date, investigations of insulin transdermal delivery have mainly focused on two other approaches to increase skin permeability to insulin: iontophoresis, the application of an electric field across intact skin, and sonophoresis, the application of ultrasound. Both are established, effective delivery systems that require further safety research and parameter optimization.

Iontophoresis. The potential for iontophoretically-aided transdermal insulin delivery depends on the successful manipulation of insulin's physicochemical properties (e.g., charge) and the electrical field employed (current, time, waveform, frequency, and on:off ratio). The weak ionization of insulin and its tendency to aggregate may have contributed to early failures of transdermal delivery in humans. After an unsuccessful preliminary attempt to deliver insulin by iontophoresis in humans (211), new insights have been gained for improving this mode of administration. For example, a modified, highly charged and monomeric form of insulin induced a hypoglycemic effect in a porcine model (211).

Insulin charge density varies with solution pH. At pH 3.7, insulin solutions are more conductive than around neutral pH, or at the isoelectric pH of 5.3 (*212*). This pH 3.7 insulin solution, when delivered iontophoretically in diabetic rats, caused a significant decrease in the blood glucose level. In an effort to curtail the time of exposure to current and the delivery time in general, the dependence of insulin delivery on pulse waveform modes, frequency, and current intensity was established (*213*). A pulsed current, with a 1:1 on:off ratio was found to be optimal, since it minimized skin polarization (which may inhibit insulin delivery). Increased pulse frequency was shown to improve insulin delivery in a diabetic hairless rat, with effects lasting past 12 hours. An electro-osmotic device tested in rabbits also caused a similar drop in glucose and duration of effect (*214*).

To increase the transdermal delivery of insulin during iontophoresis, several approaches have been used. For example, insulin has been modified chemically to increase its iontophoretic transport. One successful approach was to succinylate the hormone, which greatly increased its charge and resulted in a monomeric form which exhibited a 20-fold increase in iontophoretic transport across hairless mouse skin compared to the unmodified hormone (*215*). Another approach was to use enhancers which may affect skin permeability. The result of mixing such agents with the iontophoresed insulin resulted in up to 200% enhancement of transport through human cadaver skin (*216*). The toxicity of these enhancers, however, has yet to be determined.

Sonophoresis. Ultrasound has been shown to enhance skin permeability by inducing cavitation and creating disorder in the keratinocytes of the stratum corneum, thus allowing the passage of drugs (*217*). In hairless mice partially immersed in an insulin solution, a 5-minute application of ultrasound reduced blood sugar to 22% of its initial value (*218*). In another study, a 90-minute application of pulsed ultrasound was used to deliver insulin to diabetic rabbits; glucose levels dropped about 50% and returned to baseline in 7 hours (*219*). There were no signs of burns, erythema, inflammation, or destruction of tissue after these experiments. Finally, in hairless rats, ultrasound applied for only one hour has been shown to produce a hypoglycemic profile nearly identical to that generated following a subcutaneous injection (*220*).

It is proposed that the development of a small, pocket-size sonicator (capable of delivering a reasonable dose of insulin to a small application area in a short period of time) could make sonophoretic delivery of daily therapeutic doses of insulin feasible. More work needs to be done to evaluate the safety effects of chronic transdermal administration aided by ionto- or sonophoresis. It is anticipated that human clinical trials of an ultrasound-mediated transdermal insulin delivery device will start in the near future (*221*).

Oral Administration. Oral delivery is the most popular route of drug administration. However, the oral delivery of proteins and insulin in particular is limited by acid-catalyzed degradation in the stomach, proteolytic breakdown in the gastrointestinal tract, and poor permeability across the gastrointestinal mucosa. A convenient oral dosage form of insulin is still worth investigating, however, since it offers a physiological method of delivery. Insulin that remains bioactive in the intestinal tract can penetrate through the intestinal membrane and eventually diffuse into the portal vein. In this case, hepatic glucose production is reduced, and insulin degradation decreases the amount of peripherally circulating insulin.

Penetration through the intestinal wall occurs via several possible mechanisms. Insulin can be transported through the vast surface area of intestinal cells by endocytosis, receptor-mediated endocytosis, or facilitated transport, and then diffuse into mesenteric veins to the portal vein. Penetration through some intestinal cells (Peyer's patches) leads to lymph vessels. Insulin absorbed into the lymphatic circulation drains into the thoracic duct, which empties into the bloodstream at the subclavian vein. The amount of peptide absorption which results in lymphatic absorption is not well-documented, although the denser vascularization of veins gives reason to suspect that the bulk of the absorbed insulin goes directly to the liver. Alternatively, shredding of enterocytes has also been suggested as another method of insulin absorption. A thorough review of the mechanism of absorption (with an emphasis on microemulsions) can be found elsewhere (222).

Many aspects associated with oral delivery, including low bioavailability, variability of dosage effects and the need for absorption enhancers or protease inhibitors (of questionable safety), limit the potential of a tablet or capsule dosage form of insulin. Inconsistency of insulin absorption is the main culprit: the time of passage through the digestive tract is not predictable, especially since membrane permeabilities are variable. Many factors influence gastrointestinal absorption, including the changing surface area for absorption, the pH, and different proteolytic enzyme contents, residence times, bacteria populations, viscosity, and the number of Peyer's patches in each intestinal division (222). These complications make accurate and optimized delivery problematic. Even though the small intestine has a large surface area, targeting is favored towards the large intestine and rectum, due to their lower enzyme content and longer residence time of materials. Intestinal cells within a given region have variable properties, and may differ in permeability. This leads to variable absorption of insulin, which in turn results in a variable hypoglycemic response (223).

Approaches to oral delivery have focused on encapsulation techniques and controlled-release mechanisms for their ability to shield oral insulin from degradation in the stomach. Various protease inhibitors and absorption promoters have also been investigated in animal models to evaluate the potential of GI delivery (224-232). Capsule systems have been devised to release their contents after a certain time lag, or at a given pH (230). These delivery methods have yielded mediocre absorption and unsurprisingly variable results (2-13% bioavailability, 8-60% reduction in glucose levels).

A significant improvement in the oral absorption of insulin has been achieved by increasing its lipophilicity. For example, research has been conducted on the feasibility of liposome-encapsulated oral insulin, but extremely high doses (233,234), variability (235), and high cost of production (236) deem liposomal delivery of insulin impractical. Other approaches include solid dispersion in stearic acid (237), embedding in water-in-oil microemulsions (238), chemical modification with palmitic acid (239), and fatty acid dispersions (240,241). It was shown that saturated fatty acids were more effective than unsaturated ones, as evidenced by better control of blood glucose in a rabbit model.

Although significant effort has been dedicated towards oral insulin administration in animal models, the applicability to diabetic humans has yet to be demonstrated. Issues related to the precision of drug targeting and the safety of absorption enhancers and protease inhibitors need to be addressed. Furthermore, orally administered insulin exhibits a relatively slow uptake, making it, at best, appropriate only for maintaining a basal level.

Nevertheless, oral delivery of insulin is being pursued in the pharmaceutical industry, albeit with caution. For example, early promising results reported for insulin delivered from water-in-oil microemulsions (238) were later shown to be fraudulent as the oral insulin product was contaminated with a therapeutic dose of another diabetes drug, glibinimide (242). The sponsoring company, Cortecs Ltd. (U. K.), immediately suspended the trials. Despite the setback, some pharmaceuticals companies are still pursuing the elusive goal of delivering insulin orally. In an intriguing, related development, Eli Lilly (in partnership with Autoimmune, Lexington, MA), is in Phase II trials of an oral insulin product (243). This trial is based on oral tolerance therapy, instead of having the ingested insulin directly affecting blood glucose. The idea is based on the body's ability to invoke a protective immune response to consumed proteins; insulin processing in the gastrointestinal tract is expected to lead to regulatory T-cell formation which, in turn, may release suppressor cytokines preventing the destruction of islet cells in the pancreas. Another novel approach towards oral delivery of insulin has been reported recently which employs a partially unfolded (molten globule state) molecule exhibiting increased permeability across the intestinal epithelium (*244*). The company involved, Emisphere Technologies (Hawthorne, NY), is in the pre-clinical stages of the technology employing a murine diabetes model (*245*).

Other Delivery Routes. Efforts have been directed towards additional routes of insulin administration. While these studies have yielded less success clinically, they are nonetheless useful in studying the potential and limitations of non-parenteral protein delivery routes.

Rectal. Delivery of insulin via the rectal tract is both non-parenteral and physiological, advantages shared with oral administration. However, with the rectal route several disadvantages of oral administration are avoided. These include the high variability of absorption (due to environmental differences along the gastrointestinal tract) and dilution of the dose.

Rectally delivered insulin can be targeted to absorb into the superior hemorrhoidal vein (the vein furthest away from the anus) and consequently pass into the portal vein, which leads to first-pass metabolism (*246*). The amount of proteolytic enzymes in the rectal tract is much lower than that encountered orally, however, the permeability barrier of the membrane remains. As a result, many absorption enhancers have been tested to improve rectal insulin formulations. These include surfactants (*247*), enamines (*248-250*), bile salts (*251-253*) and their derivatives (*254*), sodium salicylate (*250,255*), and other formulations in hollow-type suppositories (*256,257*). The long-term safety and mechanism of action of these enhancers needs to be addressed. In the case of surfactants, enhanced absorption is thought to be due to inhibition of peptide degradation by mucousal enzymes, in addition to opening epithelial tight junctions and disrupting the lipid bilayer of cell membranes, which reduces the mucus layer viscosity (*258*). Certain surfactants, however, are suspected to cause lasting effects on the mucousal epithelium. For bile salts, a different mechanism has been suggested, involving solubilization of the insulin monomer and formation of reverse micelles and pores within cellular membranes, allowing diffusion through cells (*194*). Mucousal irritation has been noted with the use of bile salts (*194*).

By using the enhancers discussed above, a relatively rapid hypoglycemic effect can be achieved via rectal insulin delivery. For example, in healthy rabbits, a rectally delivered insulin dose of 0.5 U/kg or 8 U/kg lowered the glucose level by 35% and 75%, respectively, with the peak effect observed within 45 min to one hour (*250,251,256*). Similar results were obtained from studies comparing non-diabetic subjects (*254*) to type I diabetics (*259*); insulin doses of 1.5-2 U/kg delivered rectally (with sodium salicylate as enhancer) were effective in lowering plasma glucose some 60-70% in less than one hour. Another study of diabetics demonstrated a less pronounced effect using the same conditions, but a different enhancer, 3% polyoxyethylene-9-lauryl ether (*260*). In that study, some participants complained of abdominal discomfort or a feeling of rectal urgency (*260*).

The encouraging results in humans may warrant the application of rectal insulin delivery for preprandial (rapid-acting) doses. However, safety (i.e., use of enhancers) and compliance issues remain, particularly for use in the U. S. market. This approach may be promising for the European market, where the delivery of pharmaceuticals by suppository is more common.

Vaginal Delivery. Vaginal delivery of insulin has also been explored. Drug absorption through the vaginal mucosa faces limitations common to other mucousal delivery routes, including low permeability (which can be ameliorated by use of absorption enhancers (*261*)) and a proteolytic barrier similar to the ileal and rectal mucosa (*262*). However, the large surface area and rich venous plexus surrounding the vagina have fostered the potential for systemic absorption. Absorbed drug would empty into the internal iliac veins, and enter circulation without first-pass metabolism in the liver (*263*).

In order to achieve successful, predictable vaginal insulin delivery, one must understand and overcome the many factors which complicate absorption. These include changes in permeability, volume, composition, and pH of vaginal fluids and cervical mucus (*263*). For example, a study in rats showed that the estrous cycle markedly affected vaginal insulin absorption (*264*). The optimal hypoglycemic response from a 20-U dose (including citric acid as enhancer) was a drop to about 30% of initial glucose levels in 3-4 hours. This would be classified as intermediate-acting. The absolute bioavailability of this formulation was estimated to be 18% (*265*). Another study examining the effects of several different enhancers yielded similar results (*258*).

Delivery of intrauterine-instilled insulin may be another option. A study using this approach found that the menstrual cycle did not seem to affect uterine absorption (*266*). Furthermore, an enhancer was not necessary; a 0.4 U/kg dose yielded a response similar to that achieved from the same amount delivered subcutaneously (*266*).

Further research on vaginal insulin delivery is necessary in order for this approach to emerge as a treatment for diabetes. Insulin may, however, be a useful model protein in the upcoming years for the investigation of optimized vaginal or intrauterine peptide delivery.

Buccal Delivery. Less extensive work has been done on the buccal delivery of insulin, primarily due to the very low bioavailability achieved by this technique. Molecules of low atomic weight have been known to absorb quickly into the reticulated vein, entering the systemic circulation directly. The limitations to transmucousal delivery include the need for enhancers and/or protease inhibitors and the inherent variability of absorption resulting from the patient's eating and drinking habits, and tongue movements.

Experimental investigations of bucally delivered insulin have not been promising. For example, an investigation involving healthy, anesthetized dogs demonstrated that bucally delivered insulin (without an enhancer) had only 0.5% availability relative to intramuscular injection (*267*). Another study, which examined the effect of formulation variables including the addition of enhancers, also demonstrated very low buccal bioavailability in addition to highly variable delivery (*268*). Liposomal delivery and entrapment in erythrocyte ghosts did not fare significantly better with regards to hypoglycemic effect and variability (*235,269*).

Buccal delivery does not appear to be applicable for diabetes treatment since accurate delivery is a key concern. Furthermore, the oral mucosa is a very sensitive membrane. Since taste and toxicity are special considerations, substantial work is necessary to make buccal insulin delivery a reality.

Ocular Delivery. Peptides and proteins may be delivered through the ocular route (*270*), primarily for inflammation, wounds, and glaucoma. As early as 1931, Christie and Hanzal (*271*) demonstrated that insulin could be delivered ocularly, resulting in a dose-response lowering of blood glucose in rabbits.

More recent studies have investigated insulin delivery via this route, but several obstacles still persist. Insulin is susceptible to metabolism by peptidases in ocular tissues and also requires an enhancer for delivery (due to the hormone's large size) which could be a problem considering the extremely sensitive tissues surrounding the eyes. The conjunctiva, cornea, and eyeball must be compatible with any proposed enhancer, since the enhancer could stimulate absorption into any of these sites. A drug instilled to the eye is rapidly eliminated from the pre-corneal area by penetration across the cornea, conjunctiva, and sclera, or by drainage into the nasolacrimal duct. This duct is one pathway to systemic circulation (272), although experiments have shown that a fair amount of insulin absorption occurs through the conjunctiva (273,274).

An insulin dosage of approximately 4-6 U/kg with 0.5% saponin was shown to reduce glucose to 30% or 50% of initial levels in healthy rabbits and rats, respectively (275,276). The hypoglycemic effect was superior to the 20% reduction in glucose levels achieved with the same dosage given intravenously (275). However, saponin at concentrations above 0.5% causes ocular irritation, as do polyoxyethylene ether enhancers BL-9 and Brij 78 to a lesser extent (276). In healthy cats, 0.5% saponin and Brij-99 caused irritation, although solutions containing Brij-78 and BL-9 were well-tolerated (277). Also in the cat, a similar hypoglycemic effect of 50-60% of initial values was induced by applying 15 U of insulin with 0.5% saponin or 1% BL-9, although the time to the peak drop in glucose levels was only 30 min. In search for non-toxic and non-irritating enhancers, alkylglycosides, which metabolize into non-toxic products (276), DS-1, a modified saponin (278), benzalkonium chloride, and paraben (279,280) gave positive preliminary results in animal models. However, more detailed studies are necessary to characterize the effects of enhancers on ocular tissues.

The potential of ocular insulin delivery may be contingent upon the future of nasal delivery, which so far has been found to provide better absorption with tested enhancers (276,278). Human trials using enhancers for ocular delivery have not yet been thoroughly investigated.

STABILITY ISSUES CONCERNING INSULIN FORMULATIONS AND DELIVERY SYSTEMS

The discussion above regarding developments in insulin delivery illustrates that although many approaches have been investigated, few have reached clinical use. One of the primary reasons for this is that insulin, like most proteins and peptides (281), is a relatively fragile molecule. This fragility has been thoroughly studied. Due to its pharmaceutical importance, insulin has provided an excellent model for investigation. Understanding this fragility is critical towards advancing insulin's pharmaceutical use. The purpose herein is not to provide a comprehensive survey of insulin stability since this can be found elsewhere (41,282), but to present a critical overview with an emphasis on how insulin's stability impacts its pharmaceutical use.

Early Observations of Insulin Stability. Even with the original discovery of insulin by Banting and Best, it was noted that biological potency was destroyed by boiling or pancreatic juices, but not by tricresol (17). Although the concept of insulin shelf-life had not been established, it was also noted that the pancreatic extracts kept their potency for at least a week in cold storage (17). Within about a decade, researchers using more purified insulins investigated and elucidated inactivation by heat and alkaline pH (283), evolution of ammonia during treatment with acid or alkali (284), and destruction in the presence of sulfhydryl compounds (285,286). As more purified, crystalline insulin became available and clinically used, it was found that heating insulin in acidic medium induced precipitation (287-289) later shown to be related to insulin fibrillation (289). In those early days, the susceptibility of insulin towards acid- and heat-induced inactivation varied quite widely, necessitating that each lot be tested to meet certain requirements (see above).

Physical Stability (Insulin Fibrillation/Non-covalent Aggregation). The native three dimensional structure of insulin is required for receptor binding, and thus activity. This structure is lost when insulin molecules combine together to form fibrils. Such fibrils were described in the early years of insulin discovery, but it was not until a few decades later that Waugh (*290-292*) found that heat-induced fibrillation of insulin proceeded in three stages: nucleation (or the formation of active centers), growth, and precipitation. While nucleation requires an elevated temperature, the latter two stages can occur at ambient or even lower temperatures. The fibrils were shown to be aggregates of insulin molecules held together by non-covalent forces. Fibril growth was shown to depend on surface area of the aggregates, ionic strength, and the concentration of insulin in solution. Surprisingly, it was found that whereas growth rate increases with increasing concentration for acidic solution, neutral insulin solution demonstrated the reverse trend.

Today, this phenomenon is understood in terms of the association state of insulin. Under acidic conditions or at very low concentrations, insulin exists primarily as a monomer (*293,294*), i.e., the species involved in fibrillation. Neutral pH and higher concentrations favor the more stable hexameric species. The fibrillation is initiated when the insulin monomer partially unfolds (*41,282*), for example when it binds to a hydrophobic interface, presumably via the protein's hydrophobic patches normally buried upon hexamerization (*295*). Examples include the air-water interface or polymeric surfaces. Light scattering measurements of the insulin fibrillation process suggest that partially unfolded insulin monomers associate to form intermediate size aggregates. When these reach a critical size (about 170 nm, corresponding to some 100 monomers (*296,297*)), they have enough surface energy to remain stable and may associate with native insulin (*297*). Computer simulations of this model were successful in describing the kinetics (in particular the lag phase) observed for insulin aggregation.

It is noteworthy that for the first half century of insulin's clinical use, hardly any mention had been made in the literature regarding fibrillation-related stability problems during normal handling, storage, or administration (*282*). However, this issue became paramount from the late 1970s onward, with the advent of insulin infusion devices (see above). The use of insulin solutions in such a system involves exposure to elevated temperature, presence of hydrophobic surfaces, and shear forces–all factors increasing insulin's susceptibility towards aggregation. These aggregates are responsible for the loss of biological potency in a variety of delivery environments. In addition, it is the probable cause of "frosting" of refrigerated insulin solutions, which is also accompanied by marked loss of biological activity (*16*).

One strategy to ameliorate the aggregation is to modify the insulin molecule to make it inherently more stable against aggregation. To this end, researchers are attempting to identify key residues involved in the aggregation process. Some results suggest that the formation of fibrils involves the hydrophobic residues Ile^{A2}, Leu^{B11}, and Leu^{B15} (*15*). These residues are buried in the native insulin structure (*41*). It is hypothesized that during unfolding the C-terminus of the B-chain is displaced, which results in the above-mentioned residues being brought to the protein's surface, facilitating their involvement in aggregation (*15*). By altering these and perhaps other residues, it may be possible to genetically engineer an insulin molecule that is less likely to aggregate. Another stabilization strategy is to chemically derivatize it to a species less likely to bind to hydrophobic surfaces (*298-301*).

The most common approach used to stabilize insulin against aggregation is to choose a formulation in which the protein exhibits maximal stability. A wide variety of agents have been tested for this purpose (*282*), including organic co-solvents (*302,303*), as well as organic (*299,303-309*), and inorganic (*310-314*) additives. Of these, the use of anionic or non-ionic surfactants is particularly attractive (*302,303,306,308*). Sluzky and co-workers (*303*) have provided a mechanistic rationale: being surface-active, these agents should occupy sites at the air-water (or other hydrophobic) interface, thus preventing insulin from doing so and consequently enhancing its solution stability. The most potent additives identified thus far are non-ionic sugar-based detergents (*303*). Whereas insulin undergoes almost complete agitation-induced aggregation in only one day in the absence of surfactants (*297,303*), conventional detergents such as the Tweens and Pluronics can keep it stable for several days (*306,308*), and sugar-based non-ionic detergents, such as n-dodecyl-β-D-maltoside and n-octyl-β-D-glucopyranoside prevent insulin from aggregating for over a month (*303*). The successful application of such knowledge is needed to further the pharmaceutical use of insulin infusion systems.

Chemical Stability. In addition to non-covalent aggregation discussed above, insulin is also prone to a number of deleterious covalent processes. One such process is deamidation. This event is a common problem for protein pharmaceuticals and can occur in aqueous solution at either acidic or neutral conditions, with different chemical mechanisms responsible (*315*). Until recently, rapid-acting insulin was only available in acidic media (pH 2-3). During storage at 4 °C, this insulin preparation deamidates (primarily at Asn^{A21}) at about 1-2% per month (*41*), and the rate is even higher at room temperature (*282*). The mechanism for this process at pH 2-5 involves an intramolecular nucleophilic attack of the Asn^{A21} C-terminal carboxylic acid onto the side-chain amide carbonyl to form a reactive cyclic anhydride intermediate, which further reacts with either water or the N-terminal amino group (Phe^{B1}) of another insulin molecule to generate deamidated insulin or an amide-linked covalent dimer, respectively (*316-318*). In neutral solution, which is more relevant to infusion devices, insulin also deamidates, albeit at a different site, Asn^{B3}, and by a different mechanism, resulting in the formation of aspartate and isoaspartate residues (*319*). The rate of the process varies among commercially available insulin preparations. However, these insulin deamidation products have been shown to have the same *in vivo* biological activity and immunogenicity as native insulin (*41*). A more damaging covalent event occurring to insulin is cleavage between residues Thr^{A8} and Ser^{A9} which was found in insulin Zn^{2+}-insulin crystalline suspensions (*319*); chain-cleaved insulin has only 2% activity compared to the native insulin molecule (*320*).

In addition to deamidation and chain cleavage, insulin can also undergo covalent cross-linking. For example, the formation of covalent insulin dimers (CID) and high-molecular weight transformation products has been observed in various insulin preparations at 4-45 °C (*321*). These include CIDs and, for protamine-containing preparations, covalent insulin-protamine products. These processes are relatively slow compared to the hydrolytic and deamidation pathways (*319*). These degradation products may be a problem since they are potentially highly immunogenic; specific antibodies against CIDs have be identified in 30% of insulin-treated diabetics (*322*).

Stability in the Solid State (Implication for Sustained Release from Polymers). Even though insulin is susceptible to a variety of deleterious covalent and non-covalent processes in aqueous solution, this does not present a major impediment to its use in injection-based diabetes therapy. Other problems persist, however, which are unrelated to insulin's stability in parenteral formulations (such as non-physiological delivery, variable glycemic control, and uncomfortable administration leading to poor patient compliance). The next significant improvement in diabetes therapy may be some form of prolonged self-regulated controlled release. As discussed earlier, such a portable or implantable device is unlikely to contain an aqueous insulin solution, especially if it is intended to provide glycemic control over months or years (both insulin stability and the shear volume required for prolonged administration would undermine this approach). Thus, the success of novel insulin therapies relies to a great extent on the development of solid dosage forms, such as polymeric devices or powders for pulmonary delivery. In this case, it is the stability of the solid protein which is of concern.

Solid insulin exhibits many of the same pathways of destabilization as in aqueous solution. For example, Fisher and Porter (*323*) have observed deamidation and cross-linking in solid insulin crystals. In addition, lyophilized insulin is also susceptible to deamidation, as revealed by the work of Strickley and Anderson (*324,325*). As was seen for the aqueous solution at acid pH discussed above, solid-state deamidation proceeds via an intramolecular nucleophilic attack leading to a reactive cyclic anhydride intermediate resulting in deamidated monomers and Asp^{A21}-Phe^{B1} and Asp^{A21}-Gly^{A1} amide-linked dimers. The degradation rate was shown to be a function of the pH in aqueous solution prior to lyophilization, with the fastest reactions observed for powder lyophilized from pH $\leq$ 4 (due to protonation of the carboxy terminus of A21 with $pK_a \approx 4$, resulting in greater reactivity).

Insulin also undergoes moisture-induced aggregation in the lyophilized powder at neutral to basic conditions via non-covalent interactions and intermolecular disulfide-bonding (*326*). The latter pathway was hypothesized to proceed via a β-elimination step followed by thiol-catalyzed disulfide exchange. Several rational strategies were devised to inhibit the moisture-induced insulin aggregation. These included controlling the pH and moisture at optimal levels, lyophilization from acidic pH, and addition of trace amounts of Cu^{2+} to insulin solution prior to lyophilization in order to catalyze the oxidation of low-molecular-weight thiols to unreactive species.

A general approach developed towards stabilizing insulin (and other proteins) in the dry form is to trap it in a glassy matrix, for example that provided by a saccharide or polyol excipient. There are several hypotheses to explain this stabilizing effect, including reduction in molecular mobility (*327,328*), physical separation ("dilution") of protein molecules (*329,330*), and preservation of protein structure via protein-excipient interactions ("water substitution") (*331,332*). It was shown that entrapping insulin in a glassy matrix of trehalose was successful in inhibiting its solid-state covalent dimerization, presumably because this process is bimolecular and thus highly dependent on the molecules' mobility (*324,325*). Interestingly, this approach did not have as much effect on unimolecular deamidation, consistent with the notion that this process does not necessarily require molecular mobility, only some conformational flexibility. Even so, keeping insulin in a glassy state may be a general approach for stabilization, in particular since it has been proposed that intermolecular events are generally favored over intramolecular ones in the solid state due to the high degree of protein-protein contacts (*333*). The use of reducing sugars should be avoided, since they have been shown to react with amino groups of solid insulin in the dry state (*334*).

Although entrapment within a glassy matrix may impart stability to insulin in the dry state, this may not be sufficient for stability in the "wetted" state. For example, in the case of solid, amorphous insulin suspended within a polymeric controlled-release device, the dry protein will eventually become hydrated prior to its release. It has been shown that insulin's water content is highly linked to its solid-state deterioration. The processes of deamidation and covalent dimerization were both highly dependent on the water content, with maximal rates (for both the unimolecular and bimolecular event) occurring above 20 g water/100 g dry protein (*324,325*). Very similar behavior was observed for insulin's moisture-induced aggregation (*326*), i.e., the onset of insulin aggregation corresponded to the uptake of water above some 20% (w/w).

Studies of insulin solid-state degradation are critical to the development of future delivery regimes involving polymers. For example, it was reported that (bovine) insulin loaded within poly(lactide-*co*-glycolide) (PLGA) microspheres exhibited extremely slow release due to degradation of the protein inside the microspheres (*335*). PLGA is a promising, FDA-approved vehicle for the sustained release of proteins (*336*). The mechanism(s) responsible were not identified, but bioerosion of the polymer yielding acidic monomer units and decreasing the pH within the microspheres was implicated (*335*). The issue of insulin stability in PLGA and other polymers must be addressed in order to facilitate the development of insulin polymeric delivery devices.

CONCLUDING REMARKS

The discovery of insulin some seventy five years ago marked a milestone in the treatment of diabetes mellitus and in the development of biological drugs generally. Throughout its history, insulin has been the focus of intense study in academia as well as in industry. Insulin has been the subject of several major scientific advances from protein sequencing and structure determination to recombinant DNA technology and genetic engineering of proteins. Recognition of the medical importance of insulin as a biopharmaceutical has had an impact on the development of federal regulations of protein drugs, and, consequently, strict regulations have been passed regarding insulin products. Insulin formulations have evolved from the early days of impure extracts to highly purified preparations of recombinantly engineered human protcin. Ongoing work in the pharmaceutical development of insulin includes genetic engineering to design more efficacious molecules and investigation of alternate (i.e., non-parenteral) delivery routes such as insulin infusion systems and controlled-release from polymers.

Literature Cited

1. Orci, L.; Vassalli, J. D.; Perrelet, A. *Sci. Am.* **1988**, *259(3)*, 85-94.
2. Chien, Y. W. *Drug Devel. Ind. Pharm.* **1996**, *22*, 753-789.
3. Brange, J.; Langkjær, L. In *Stability and Characterization of Protein and Peptide Drugs: Case Histories*; Wang, Y. J.; Pearlman, R., Eds.; Plenum Press: New York, 1993; pp 315-350.
4. Derewenda, U.; Derewenda, Z.; Dodson, G. G.; Hubbard, R. E.; Korber, F. *Brit. Med. Bull.* **1989**, *45(1)*, 4-18.
5. Steiner, D. F.; Oyer, P. G. *Proc. Natl. Acad. Sci. USA* **1967**, *57*, 473-480.
6. *Diabetes Mellitus: Theory and Practice*; Elenberg, M.; Rifkin, H., Eds.; McGraw-Hill: New York, 1970.
7. McCarrren, M. *Diabetes Forecast* **1995**, *48*, 17-19.
8. The Diabetes Control and Complications Trial Research Group. *New Engl. J. Med.* **1993**, *329*, 977-986.
9. Stinson, S. C. *Chem. Engr. News* **1991**, *69(39)*, 35-59.
10. *Nonprescription Products: Formulations & Features '96-97*; Knodel, L. C.; Kendall, S. C.; Young, L. L.; Hickey, M. J., Eds.; American Pharmaceutical Association: Washington, D. C., 1996; pp 172-176.
11. Berson, S. A.; Yalow, R. S.; Bauman, A.; Rothschild, M. A.; Newerly, K. *J. Clin. Invest.* **1956**, *35*, 170.
12. Guyton, A. C. *Textbook of Medical Physiology*; W. B. Saunders Co.: Philadelphia, PA, 1986; pp 929-930.
13. Home, P. D.; Thow, J. C.; Turnbridge, F. K. E. *Brit. Med. Bull.* **1989**, *45*, 92-110.
14. Campbell, R. K.; White, J. R.; *New Opportunities in Diabetes Management*; Miles, Inc.: Terrytown, NY, 1993; pp 15.

15. Brange, J.; Hansen, J. F.; Havelund, S.; Melberg, S. G. In *Advanced Models for Therapy of Insulin-Dependent Diabetes*; Brunetti, P.; Waldhäusl, W., Eds.; Raven Press: New York, 1987; pp 85-90.
16. Benson, E. A.; Benson, J. W.; Fredlund, P. N.; Mecklenburg, R. S.; Metz, R. *Diabetes Care* **1988**, *11*, 563-566.
17. Banting, F. G.; Best, C. H. *J. Lab. Clin. Med.* **1922**, *7*, 464-472.
18. Bliss, M. In *Insulin. Its Receptor and Diabetes*; Hollenber, M. D., Ed.; Marcel Dekker: New York, 1985, pp 7-19.
19. Abel, J. J. *Proc. Natl. Acad. Sci. USA* **1926**, *12*, 132-136.
20. Scott, D. A. *Biochem. J.* **1934**, *28*, 1592-1602.
21. Hagedorn, H. C.; Jensen, B. N.; Krarup, N. B.; Wodstrup, I. *J. Am. Med. Assoc.* **1936**, *106*, 177-180.
22. Hagedorn, H. C. *Rep. Steno. Hosp.* **1946**, *1*, 25-28.
23. Krayenbühl, C.; Rosenberg, T. *Rep. Steno. Mem. Hosp. Nord. Insulinlab* **1946**, *1*, 60-73.
24. Hallas-Møller, K.; Petersen, K.; Schlichtkrull, J. *Ugeskr Laeger* **1951**, *113*, 1761-1767.
25. Hallas-Møller, K. *Diabetes* **1956**, *5*, 7-14.
26. Hutt, P. B. *Clin. Pharmacol. Ther.* **1983**, *33(4)*, 537-548.
27. U. S. Statutes **1848**, *9*, 237.
28. U. S. Statutes **1906**, *34*, 768.
29. U. S. Statutes **1902**, *32*, 728.
30. U. S. Statutes **1938**, *52*, 1040.
31. U. S. Statutes **1962**, *76*, 780.
32. U. S. Statutes **1941**, *55*, 851.
33. U. S. Statutes **1945**, *59*, 463.
34. U. S. Statutes **1947**, *61*, 11.
35. U. S. Statutes **1949**, *63*, 409.
36. U. S. Statutes **1953**, *67*, 389.
37. Holzman, D. C. *Genetic Engr. News* **1996**, *15(21)*, 1-3.
38. Johnson, J. D. In *Biopharmaceutical Statistics for Drug Development*; Peace, K. E., Ed.; Marcel Dekker: New York, 1988; pp 1-20.
39. Jorpes, J. E. *Arch. Intern. Med.* **1949**, *83*, 363-371.
40. Schlichtkrull, J.; Brange, J.; Ege, H.; Hallund, O.; Heding, L. G.; Jørgensen, K.; Markussen, J.; Stahnke, P.; Sundby, F.; Vølund, A. *Diabetologia* **1970**, *6*, 80-81.
41. Brange, J. *Galenics of Insulin*; Springer-Verlag: Berlin, 1987.
42. Jørgensen, K. H.; Hallund, O.; Heding, L. G.; Tronier, B.; Falholt, K.; Damgaard, U.; Thim, L.; Brange, J. In *Hormone Drugs*; Bransome, E. D.; Outschoorn, A. S., Eds.; United States Pharmacopeial Convention: Rockville, Maryland, 1982; pp 139-147.
43. Heine, R. J. In *World Book of Diabetes in Practice*; Krall, L. P., Ed.; Elsevier Science Publishers BV (Biomedical Division): New York, 1988, Vol. 3; pp 150-159.
44. Schlichtkrull, J. In *Diabetes Mellitus III*; Oberdisse, K.; Jahnke, K., Eds.; Kongress der International Diabetes Federation: Düsseldorf, 1959; pp 773-777.
45. Schlichtkrull, J.; Munck, O.; Jersild, M. *Acta Med. Scand.* **1965**, *177*, 103-113.
46. Umbar, F.; Störring, F. K.; Föllmer, W. *Klin. Wochenschr.* **1938**, *17*, 443-446.
47. Goerz, G.; Ruzicka, T.; Hofmann, N.; Drost, H.; Grüneklee, D. *Hautarzt* **1981**, *32*, 187-190.
48. Wintersteiner, O.; Vigneaud, V.; Jensen, H. *J. Pharmacol. Exp. Ther.* **1929**, *32*, 397-411.
49. Ryle, A. P.; Sanger, F.; Smith, L. F.; Kitai, R. *Biochem. J.* **1955**, *60*, 541-556.

50. Adams, M. J.; Blundell, T. L.; Dodson, E. J.; Dodson, G. G.; Vijayan, M.; Baker, E. N.; Harding, M. M.; Hodkin, D. C.; Rimmer, B.; Sheat, S. *Nature* **1969**, *224*, 491-495.
51. Blundell, T. L.; Cutfield, J. F.; Cutfield, S. M.; Dodson, E. J.; Dodson, G. G.; Hodgkin, D. C.; Mercola, D. A.; Vijayan, M. *Nature* **1971**, *231*, 506-511.
52. Dodson, E. J.; Dodson, G. G.; Hodgkin, D. C.; Reynolds, C. D. *Can. J. Biochem.* **1979**, *57*, 469-479.
53. Bently, G.; Dodson, E.; Dodson, G.; Hodgkin, D.; Mercola, D. *Nature* **1976**, *261*, 166-168.
54. Chothia, C.; Lesk, A. M.; Dodson, G. G.; Hodgkin, D. C. *Nature* **1983**, *302*, 500-505.
55. Dodson, E. J.; Dodson, G. G.; Lewitova, A.; Sabesan, M. *J. Mol. Biol.* **1978**, *125*, 387-396.
56. Chawdhury, S. A.; Dodson, E. J.; Dodson, G. G.; Reynolds, C. D.; Tolley, S. P.; Blundell, T. L.; Cleasby, A.; Pitts, J. E.; Wood, S. P. *Diabetologia* **1983**, *25*, 460-464.
57. Smith, G. D.; Swenson, D. C.; Dodson, E. J.; Dodson, G. G.; Reynolds, C. D. *Proc. Natl. Acad. Sci. USA* **1984**, *81*, 7093-7097.
58. Weiss, M. A.; Nguyen, D. T.; Khait, I.; Inouye, K.; Frank, B. H.; Beckage, M.; O'Shea, E.; Scoclson, S. E.; Karplus, M.; Neuringer, L. J. *Biochemistry* **1989**, *28*, 9855-9873.
59. Kline, A. D.; Justice, R. M. J. *Biochemistry* **1990**, *29*, 2906-2913.
60. Horuk, R.; Blundell, T. L.; Lazarus, N. R.; Neville, R. W. J.; Stone, D.; Wollmer, A. *Nature* **1980**, *286*, 822-824.
61. Ebine, Y.; Leland, E.; Jarnagin, K. *Cell* **1985**, *40*, 747-758.
62. Ullrich, A.; Bell, J. R.; Chen, E. Y. *Nature* **1985**, *313*, 756-761.
63. Hubbard, S. R.; Wei, L.; Ellis, L.; Hendrickson, W. A. *Nature* **1994**, *372*, 746-753.
64. Joost, H. G. *Cell-Signal.* **1995**, *7*, 85-91.
65. Blundell, T. L.; Dodson, G. G.; Hogkin, D.; Mercola, D. *Adv. Protein Chem.* **1972**, *26*, 279-402.
66. Pullen, R. A.; Lindsay, D. G.; Wood, S. P.; Tickle, I. L.; Blundell, T. L.; Wollmer, A.; Krail, G.; Brandenburg, D.; Zahn, H.; Gliemann, J.; Gammeltoft, S. *Nature* **1976**, *259*, 369-373.
67. Hallas-Møller, K. *Chemical and Biological Insulin Studies I and II (Dissertation)*; Copenhagen University: Copenhagen, 1945.
68. Kumar, D. *Diabetes* **1979**, *28*, 994-1000.
69. Steiner, D. F.; Cunningham, D.; Spigelman, L.; Aten, B. *Science* **1967**, *157*, 697-700.
70. Steiner, D. F.; Kemmler, W.; Howard, S.; Peterson, T.; Peterson, J. *Federation Proc.* **1974**, *33*, 2105-2115.
71. Steiner, D. F. *Diabetes* **1978**, *27* [Suppl 1], 145-148.
72. Frank, B. H.; Pettee, J. M.; Zimmerman, R.; Burck, P. J. In *Peptides: Synthesis-Structure-Function*; Rich, D. H.; Gross, E., Eds.; Pierce Chemical Company: Rockford, IL, 1981; pp 729-738.
73. Galloway, J. A. Hooper, S. A.; Spradlin, C. T.; Howey, D. C.; Frank, B. H.; Bowsher, R. R.; Anderson, J. H. *Diabetes Care* **1992**, *15*, 666-692.
74. Hagen, C.; Faber, O. K.; Binder, C.; Alberti, K. G. M. M. *Acta Endocrinol.* **1977**, *85* [Suppl 209], 29A.
75. Hoogwerf, B. J.; Bantle, J. P.; Gaenslen, H. E. *Metabolism* **1986**, *35*, 122-125.
76. Wahren, J.; Johansson, B.-L.; Wallberg-Heriksson, H. *Diabetologia* **1994**, *37* [Suppl 2], 99-107.
77. Brogden, R. N.; Heel, R. C. *Drugs* **1987**, *34*, 350-371.

78. Sieber, P.; Kamber, B.; Hartmann, A.; Jöhl, A.; Rinker, B.; Rittel W. *Helv. Chim. Acta* **1974**, *57*, 2617-2621.
79. Sieber, P.; Kamber, B.; Hartmann, A.; Jöhl, A.; Rinker, B.; Rittel, W. *Helv. Chim. Acta* **1977**, *60*, 27-37.
80. Markussen, J.; Damgaard, U.; Jørgensen, K. H.; Rasmussen, E.; Snel, L.; Thim, L.; Voigt, H. O. In *Hormone Drugs*; Gueriguian, J. L.; Bransome, E. D.; Outschoon, A. S., Eds.; United States Pharmacopeial Convention: Rockville, Maryland, 182 ; pp 116-126.
81. Hall, S. S. *Invisible Frontiers. The Race to Synthsize a Human Gene*; Siggwick and Jackson: London, 1988.
82. Chance, R. E.; Kroeff, E. P.; Hoffman, J. A. *Diabetes Care* **1981**, *4*, 147-154.
83. Galloway, J. A.; Anderson, J. H.; Spradlin, C. T. In *Diabetes 1988*; Larkins, R. G.; Zimmet, P. Z.; Chisholm, D. J., Eds.; Elsevier: Amsterdam, 1989; pp 85-88.
84. Markussen, J.; Damgaard, U.; Diers, I.; Fiil, N.; Hansen, M. T.; Larsen, P.; Norris, F.; Norris, K.; Snel, L.; Thim, L.; Voigt, H. O. *Diabetologia* **1986**, *29*, 568-569.
85. Thim, L.; Hansen, M. T.; Norris, K.; Hoegh, I.; Boel, E.; Forstrom, J.; Ammerer, G.; Fiil, N. P. *Proc. Natl. Acad. Sci. USA* **1986**, *83*, 6766-6770.
86. Brange, J.; Ribel, U.; Unsen, J. F.; Dodson, G.; Hansen, M. T.; Havelund, S.; Melberg, S. G.; Norris, F.; Norris, K.; Snel, L.; Sørensen, A. R.; Voigt, H. O. *Nature* **1988**, *333*, 679-682.
87. Kaarsholm, N. C.; Ludvigsen, S. *Receptor* **1995**, *5*, 1-8.
88. Brange, J.; Owens, D. R.; Kang, S.; Volund, A. *Diabetes Care* **1990**, *13*, 923-954.
89. DiMarchi, R. D.; Mayer, J.; Fan, L.; Brems, D. N.; Frank, B. H.; Green, L. K.; Hoffman, J. A.; Howey, D. C.; Long, H. B.; Shaw, W. N.; Shields, J. E.; Slieker, L. J.; Su, K. S. E.; Sundell, K. L.; Chance, R. E. In *Peptides XII*; Smith, J. A.; Rivier, J. E., Eds.; Proceedings of the 12th American Peptide Symposium; Escom: Leiden, 1992; pp 26-28.
90. Long, H. B.; Baker, J. C.; Belagaje, R. M.; DiMarchi, R. D.; Frank, B. H.; Green, L. K.; Hoffman, J. A.; Muth, W. L.; Pekar, A. H.; Reams, S. G.; Shaw,W. N.; Shields, J. E.; Slieker, L. J.; Su, K. S. E.; Sundell, K. L.; Chance, R. E. In *Peptides XII*; Smith, J. A.; Rivier, J. E., Eds.; Proceedings of the 12th American Peptide Symposium; Escom: Leiden, 1992; pp 88-90.
91. Galloway, J. A.; Chance, R. E.; Su, K. S. E. In *The Clinical Pharmacology of Biotechnology Products*; Reidenberg, M. M., Ed.; Elsevier: Amsterdam, 1991, pp 23-34.
92. Howey, D. C.; Bowsher, R. R.; Brunelle, R. L.; Woodworth, J. R. *Diabetes* **1994**, *43*, 396-402.
93. Vignati, L.; Anderson, J. H.; Brunelle, R. L.; Jefferson, F. L.; Richardson, M. *Diabetologia* **1994**, *37*, A78.
94. *Clin. Investigator News*, **1996**, *4(7)*, 33.
95. Youngman, K. M.; Bakaysa, D. L.; Kilcomons, M. A.; DeFelippis, M. R. Abstract of Papers, 211th National Meeting of the American Chemical Society, New Orleans, LA; American Chemical Society: Washington, DC, 1996; BIOT 180.
96. Bakaysa, D. L.; Frank, B. H.; Youngman, K. M.; DeFelippis, M. R. Abstract of Papers, 211th National Meeting of the American Chemical Society, New Orleans, LA; American Chemical Society: Washington, DC, 1996; BIOT 182.
97. Haycock, P. *Clin. Diabetes* **1986**, *4*, 97-118.
98. Stewart, N. L.; Darlow, B. A. *Clinical Practice* **1994**, *80*, 802-805.
99. Haffner, S. M.; Stern, M. P.; Hazuda, H. P.; Mitchell, B. D.; Patterson, J. K. *J. Am. Med. Assoc.* **1990**, *263*, 2893-2898.

100. Galloway, J. A.; Chance, R. E. *Horm. Metab. Res.* **1994**, *26*, 591-598.
101. Kerum, G.; Profozic, V.; Granic, M.; Skrabalo, Z. *Horm. Metab. Res.* **1987**, *19*, 422-425.
102. Denne, J. R.; Andrews, K. L.; Lees, D.; Mook, W. *Diabetes Educ.* **1992**, *18*, 223-227.
103. Jovanovic-Peterson, L.; Palmer, J. P.; Sparks, S.; Peterson, C. M. *Diabetes Care* **1993**, *16*, 1479-1484.
104. Lindmayer, I.; Menassa, K.; Lambert, J.; Moghrabi, A.; Legendre, L.; Legault, C.; Letendre, M.; Halle, J. P. *Diabetes Care* **1986**, *9*, 294-297.
105. Katoulis, E. C.; Raptis, S. A. *Int. J. Artif. Organs.* **1995**, *18*, 177-180.
106. Price, J. P.; Kruger, D. F.; Saravolatz, L. D.; Whitehouse, F.W. *Am. J. Infect. Control.* **1989**, *17*, 258-263.
107. Kesson, C. M.; Bailie, G. R. *Diabetes Care* **1981**, *4*, 333.
108. Ginsberg, B. H.; Parkes, J. L.; Sparacino, C. *Horm. Metab. Res.* **1994**, *26*, 584-587.
109. Albisser, A. M.; Leibel, B. S.; Ewart, T. G.; Davidovac, Z.; Botz, C. K.; Zingg, W. *Diabetes* **1974**, *23*, 389-396.
110. Pfeiffer, E. F.; Thum, C.; Clemens, A. H. *Horm. Metab. Res.* **1974**, *6*, 339-342.
111. Slama, G.; Hautecouverture, M.; Assan, R.; Tchobroutsky, G. *Diabetes* **1974**, *23*, 732-738.
112. Pickup, J. C.; Keen, H.; Viberti, G. C.; White, M. C.; Kohner, E. M.; Parsons, J. A.; Alberti, K. G. *Diabetes Care* **1980**, *2*, 290-300.
113. Strowig, S. M. *Diabetes Educ.* **1993**, *19*, 50-58.
114. Mecklenburg, R. S.; Benson, E. A.; Benson, J. W.; Blumenstein, B. A.; Fredlund, P. N.; Guinn, T. S.; Metz, R. J.; Nielsen, R. L. *New Engl. J. Med.* **1985**, *313*, 465-469.
115. Kerner, W.; Navascues, I.; Torres, A. A.; Pfeiffer, E. F. *Metabolism* **1984**, *33*, 458-464.
116. Peden, N. R.; Braaten, J. T.; McKendry, J. B. R. *Diabetes Care* **1984**, *7*, 1-5.
117. Selam, J. L.; Charles, M. A. *Diabetes Care* **1990**, *13*, 955-979.
118. Steindel, B. S.; Roe, T. R.; Costin, G.; Carlson, M.; Kaufman, F. R. *Diabetes Res. Clin. Pract.* **1995**, *27*, 199-204.
119. Dunn, F. L.; Pietri, A.; Raskin, P. *Ann. Int. Med.* **1981**, *95*, 426-431.
120. Jennings, A. M.; Lewis, K. S.; Murdoch, S.; Talbot, J. F.; Bradley, C.; Ward, J. D. *Diabetes Care* **1991**, *14*, 738-744.
121. Raskin, P.; Unger, R. H. *New Engl. J. Med.* **1978**, *299*, 433-436.
122. Pietri, A.; Ehle, A. L.; Raskin, P. *Diabetes* **1980**, *29*, 668-671.
123. Haakens, K.; Hanssen, K. F.; Dahl-Jorgensen, K.; Vaaler, S.; Aagenæs, O.; Mosand, R. *J. Int. Med.* **1990**, *228*, 457-464.
124. Hepp, K. D. *Diabetologia* **1994**, *37* [Suppl 2], S108-S111.
125. Giacca, A.; Caumo, A.; Galimberti, G.; Petrella, G.; Librenti, M. C.; Scavini, M.; Pozza, G.; Micossi, P. *J. Clin. Endocrinol. Metab.* **1993**, *77*, 738-742.
126. Hermans, M. P.; van Ypersele de Strihou, M.; Ketelslegers, J. M.; Squifflet, J. P.; Buysschaert, M. *Transplant Proc.* **1995**, *27*, 3329-3330.
127. Selam, J. L.; Kashyap, M.; Alberti, K. G. M. M.; Lozano, J.; Hanna, M.; Turner, D.; Jeandidier, N.; Chan, E.; Charles, M. A. *Metabolism* **1989**, *38*, 908-912.
128. Duckworth, W. C.; Saudek, C. D.; Henry, R. R. *Diabetes* **1992**, *41*, 657-661.
129. The Point Study Group. *Lancet* **1988**, *8616*, 866-869.
130. Saudek, C. D.; Selam, J. L.; Pitt, H. A.; Waxman, K.; Rubio, M.; Jeandidier, N.; Turner, D.; Fischell, R. E.; Charles, M. A. *New Engl. J. Med.* **1989**, *321*, 574-579.

131. Hanaire-Broutin, H.; Broussolle, C.; Jeandidier, N.; Renard, E.; Guerci, B.; Haardt, M. J.; Lassmann-Vague, V. *Diabetes Care* **1995**, *18*, 388-392.
132. Olsen, C. L.; Liu, G.; Iravani, M.; Nguyen, S.; Khourdadjuan, K.; Turner, D. S.; Waxman, K.; Selam, J. L.; Charles, M. A. *Int. J. Artif. Organs* **1993**, *16*, 847-854.
133. Meize-Grochowski, A. R. *Diabetes Educ.* **1989**, *15*, 50-55.
134. Renard, E.; Bringer, J.; Jacques-Apostol, D.; Lauton, D.; Mestre, C.; Costalat, G.; Jaffiol, C. *Diabetes Care* **1994**, *17*, 1064-1066.
135. Renard, E.; Baldet, P.; Pico, M. C.; Jacques-Apostol, D.; Lauton, D.; Costalat, G.; Bringer, J.; Jaffiol, C. *Diabetes Care* **1995**, *18*, 300-306.
136. Olsen, C. L.; Chan, E.; Turner, D. S.; Iravani, M.; Nagy, M.; Selam, J. L.; Wong, N. D.; Waxman, K.; Charles, M. A. *Diabetes Care* **1994**, *17*, 169-176.
137. Jeandidier, N.; Boivin, S.; Sapin, R.; Rosart-Ortega, F.; Uring-Lambert, B.; Reville, P.; Pinget, M. *Diabetologia* **1995**, *38*, 577-584.
138. Gombotz, W. R.; Pettit, D. K. *Bioconj. Chem.* **1995**, *6*, 332-351.
139. Leong, K.W.; Kost, J.; Mathiowitz, E.; Langer, R. *Biomaterials* **1986**, *7*, 364-371.
140. Matiowitz, E.; Langer, R. *J. Control. Release* **1987**, *5*, 13-22.
141. Kwong, A. K.; Chou, S.; Sun, A. M.; Sefton, M. V.; Goosen, M. F. A. *J. Control. Release* **1986**, *4*, 47-62.
142. Yamakawa, I.; Kawahara, M.; Watanabe, S.; Miyake, Y. *J. Pharm. Sci.* **1990**, *79*, 505-509.
143. Creque, H. M.; Langer, R.; Folkman, J. *Diabetes* **1980**, *29*, 37-40.
144. Brown, L.; Siemer, L.; Munoz, C.; Langer, R. *Diabetes* **1986**, *35*, 684-691.
145. Brown, L.; Munoz, C.; Siemer, L.; Edelman, E.; Langer, R. *Diabetes* **1986**, *35*, 692-697.
146. Arar, H. H.; Bajpai, P. K. *Biomed. Sci. Instrum.* **1992**, *28*, 173-178.
147. Atkins, T. W.; McCallion, R. L.; Tighe, B. J. *J. Biomed. Mat. Res.* **1995**, *29*, 291-298.
148. Goosen, M. F. A.; Leung, Y. F.; O'Shea, G. M.; Chou, S.; Sun, A. M. *Diabetes* **1983**, *32*, 478-481.
149. Shinde, B. G.; Nithianandam, V. S.; Kaleem, K.; Erhan, S. *Biomed. Mater. Engr.* **1992**, *2*, 123-136.
150. Shinde, B. G.; Erhan, S. *Biomed. Mater. Engr.* **1992**, *2*, 127-131
151. Kost, J.; Wolfrum, J.; Langer, R. *J. Biomed. Mat. Res.* **1987**, *21*, 1367-1373.
152. Saslawski, O.; Weingarten, C.; Benoit, J. P.; Couvreur, P. *Life Sci.* **1988**, *42*, 1521-1528.
153. Miyazaki, S.; Yokouchi, C.; Takada, M. *J. Pharm. Pharmacol.* **1988**, *40*, 716-717.
154. Wang, P. Y. *Biomaterials* **1989**, *10*, 197-201.
155. Wang, P. Y. *J. Biomed. Engr.* **1993**, *15*, 106-112.
156. Fischel-Ghodsian, F.; Brown, L.; Mathiowitz, E.; Brandenburg, D.; Langer, R. *Proc. Natl. Acad. Sci. USA* **1988**, *85*, 2403-2406.
157. Fischel-Ghodsian, F.; Newton, J. M. *J. Drug Target.* **1993**, *1*, 67-80.
158. Kim, C.-K.; Im, E.-B.; Lim, S.-J.; Oh, Y.-K.; Han, S.-K. *Int. J. Pharmaceutics* **1994**, *101*, 191-197.
159. Heller, J.; Chang, A. C.; Rodd, G.; Grodsky, G. M. *J. Control. Release* **1990**, *13*, 295-302.
160. Kost, J.; Langer, R. *Trends Biotechnol.* **1992**, *10*, 127-131.
161. Ito, Y.; Casolaro, M.; Kono, K.; Imanishi, Y. *J. Control. Rel.* **1989**, *10*, 195-203.
162. Ito, Y.; Chung, D. J.; Imanishi, Y. *Bioconj. Chem.* **1994**, *5*, 84-87.

163. Podual, K.; Doyle, F. J.; Peppas, N. A. Abstract of Papers, 211th National Meeting of the American Chemical Society, New Orleans, LA; American Chemical Society: Washington, DC, 1996; PMSE 221.
164. Armour, J. C.; Lucisano, J. Y.; McKean, B. D.; Gough, D. A. *Diabetes* **1990**, *39*, 1519-1526.
165. Creighton, T. E. *Proteins. Structures and Molecular Properties*; W. H. Freeman and Co.: New York, 1984; pp 381-382.
166. Brownlee, M.; Cerami, A. *Science* **1979**, *206*, 1190-1191.
167. Brownlee, M.;Cerami, A. *Diabetes* **1983**, *32*, 499-504.
168. Jeong, S. Y.; Kim, S. W.; Holmberg, D. L.; McRea, J. C. *J. Control. Rel.* **1985**, *2*, 143-152.
169. Seminoff, L. A.; Gleeson, J. M.; Zheng, J.; Olsen, G.; Holmberg, D.; Mohammad, S. F.; Wilson, D.; Kim, S.W. *Int. J. Pharm.* **1989**, *54*, 251-257.
170. Sato, S.; Jeong, S. Y.; McRea, J. C.; Kim, S. W. *J. Control. Release* **1984**, *1*, 67-77.
171. Kim, S. W.; Pai, C. M.; Makino, K.; Seminoff, L. A.; Holmberg, D. L.; Gleeson, J. M.; Wilson, D. E.; Mack, E. J. *J. Control. Rel.* **1990**, *11*, 193-201.
172. Baudys, M.; Uchio, T.; Mix, D.; Wilson, D.; Kim, S. W. *J. Pharm. Sci.* **1995**, *84*, 28-33.
173. Pai, C. M.; Bae, Y. H.; Mack, E. J.; Wilson, D. E.; Kim, S. W. *J. Pharm. Sci.* **1992**, *81*, 532-536.
174. Tomioka, K.; Fukuda, H.; Taniguchi, H. *J. Ferment. Bioeng.* **1994**, *77*, 442-444.
175. Kitano, S.; Koyama, Y.; Kataoka, K.; Okano, T.; Sakurai, Y. *J. Control. Rel.* **1992**, *19*, 162-170.
176. Goosen, M. F. A.; O'Shea, G. M.; Gharapetian, H. M.; Chou, S.; Sun, A. M. *Biotechnol. Bioeng.* **1985**, *27*, 146-150.
177. Lau, H. T.; Yu, M.; Fontana, A.; Stoeckert, C. J. *Science* **1996**, *273*, 109-112.
178. Illum, L.; Davis, S. S. *Clin. Pharmacokinet.* **1992**, *23*, 30-41.
179. Tengamnuay, P.; Mitra, A. K. *Pharm. Res.* **1990**, *7*, 370-375.
180. Pontiroli, A. E.; Alberetto, M.; Secchi, A.; Dossi, G.; Bosi, I.; Pozza, G. *Brit. Med. J.* **1982**, *284*, 303-306.
181. Moses, A. C.; Gordon, G. S.; Carey, M. C.; Flier, J. S. *Diabetes* **1983**, *32*, 1040-1047.
182. Salzman, R.; Manson, J. E.; Griffing, G. T.; Kimmerle, R.; Ruderman, N.; McCall, A.; Stoltz, E. I.; Mullin, C.; Small, D.; Armstrong, J.; Melby, J. C. *New Engl. J. Med.* **1985**, *312*, 1078-1084.
183. el-Etr, M.; Slama, G.; Desplanque, N. *Lancet* **1987**, *8567*, 1085-1086.
184. Frauman, A. G.; Cooper, M. E.; Parsons, B. J.; Jerums. G.; Louis, W. J. *Diabetes Care* **1987**, *10*, 573-578.
185. Frauman, A. G.; Jerums, G.; Louis, W. J. *Diab. Res. Clin. Pract.* **1987**, *3*, 197-202.
186. Pontiroli, A. E.; Calderara, A.; Pozza, G. *Clin. Pharmacokinet.* **1989**, *17*, 299-307.
187. Lassmann-Vague, V., Thiers, D.; Vialettes, B.; Vague, P. H. *Lancet* **1988**, *8581*, 367-368.
188. Sinay, I. R.; Schlimovich, S.; Damilano, S.; Cagide, A. L.; Faingold, M. C.; Facco, E. B.; Gurfinkiel, M. S.; Arias, P. *Horm. Meta. Res.* **1990**, *22*, 307-308.
189. Nolte, M. S.; Taboga, C.; Salamon, E.; Moses, A.; Longenecker, J.; Flier, J.; Karam, J. H. *Horm. Meta. Res.* **1990**, *22*, 170-174.
190. Bruce, D. G.; Chisholm, D. J.; Storlien, L. H.; Borkman, M.; Kraegen, E. W. *Diabet. Med.* **1991**, *8*, 366-370.
191. Jacobs, M. A. J. M.; Schreuder, R. H.; Jap-a-joe, K.; Nauta, J. J.; Andersen, P. M.; Heine, R. J. *Diabetes* **1993**, *42*, 1649-1655.

192. Coates, P. A.; Ismail, I. S.; Luzio, S. D.; Griffiths, I.; Ollerton, R. L.; Volund, A.; Owens, D. R. *Diabet. Med.* **1995**, *12*, 235-239.
193. Bolli, G.; DeFeo, P.; Compagnucci, P.; Cartechini, M. G.; Angeletti, G.; Santeusanio, F.; Brunetti, P.; Gerich, J. E. *Diabetes* **1983**, *32*, 134-141.
194. Gordon, G. S.; Moses, A. C.; Silver, R. D.; Flier, J. S.; Carey, M. C. *Proc. Natl. Acad. Sci. USA* **1985**, *82*, 7419-7423.
195. Merkus, F. W. H. M.; Verhoef, J. C.; Romeijn, S. G.; Schipper, N. G. *Pharm. Res.* **1991**, *8*, 588-592.
196. Merkus, F. W. H. M.; Verhoef, J. C.; Romeijn, S. G.; Schipper, N. G. M. *Pharm. Res.* **1991**, *8*, 1343.
197. Shao, Z.; Krishnamoorthy, R.; Mitra, A. K. *Pharm. Res.* **1992**, *9*, 1157-1163.
198. Björk, E.; Edman, P. *Int. J. Pharm.* **1988**, *47*, 233-238.
199. Mishima, M.; Wakita,Y.; Nakano, M. *J. Pharmacobio.-Dyn.* **1987**, *10*, 624-631.
200. Longenecker, J. P.; Moses, A. C.; Flier, J. S.; Silver, R. D.; Carey, M. C.; Dubovi, E. J. *J. Pharm. Sci.* **1987**, *76*, 351-355.
201. Dondeti, P.; Zia, H.; Needham, T. E. *Int. J. Pharm.* **1995**, *122*, 91-105.
202. Newman, S. P.; Steed, K. P.; Hardy, J. G.; Wilding, I. R.; Hooper, G.; Sparrow, R. A. *J. Pharm. Pharmacol.* **1994**, *46*, 657-660.
203. Liu, F. Y.; Kildsig, D. O.; Mitra, A. K. *Life Sci.* **1992**, *51*,1683-1689.
204. Colthorpe, P.; Farr, S. J.; Taylor, G.; Smith, I. J.; Wyatt, D. *Pharm. Res.* **1992**, *9*, 764-768.
205. Yamamoto, A.; Umemori, S.; Muranishi, S. *J. Pharm. Pharmacol.* **1994**, *46*, 14-18.
206. Liu, F. Y.; Shao, Z.; Kildsig, D. O.; Mitra, A. K. *Pharm. Res.* **1993**, *10*, 228-232.
207. Laube, B. L.; Georgopoulos, A.; Adams, G. K. *JAMA* **1993**, *269*, 2106-2109.
208. *Clin. Investigator News* **1996**, *4(4)*, 12.
209. Bromberg, L.; Klibanov, A. M. *Proc. Natl. Acad. Sci. USA* **1995**, *92*, 1262-1266.
210. Blume, G.; Cevc, G.; Schätzlein, A. *Procced. Intern. Symp. Control. Rel. Bioact. Mater.* **1996**, *23*, 713-714.
211. Stephen, N. L.; Petelenz, T. J.; Jacobsen, S. C. *Biomed. Biochim. Acta* **1984**, *43*, 553-558.
212. Siddiqui, O.; Sun, Y.; Liu, J. C. Chien, Y. W. *J. Pharm. Sci.* **1987**, *76*, 341-345.
213. Liu, J. C.; Sun, Y.; Siddiqui, O.; Chien, Y. W.; Shi, W. M.; Li, J. *Int. J. Pharm.* **1988**, *44*, 197-204.
214. Meyer, B. R.; Katzeff, H. L.; Eschback, J. C.; Trimmer, J.; Zacharias, S. B.; Rosen, S.; Sibalis, D. *Am. J. Med. Sci.* **1989**, *297*, 321-325.
215. Corbett, J. T.; Zimmerman, J. K.; Michniak, B. B.; Shalaby, S. W. Abstract of Papers, 211th National Meeting of the American Chemical Society, New Orleans, LA; American Chemical Society: Washington, DC, 1996; BTEC 007.
216. Rao, V. U.; Misra, A. N. *Pharmazie* **1994**, *49*, 538-539.
217. Mitragotri, S.; Edwards, W.; Blankschtein, D.; Langer, R. *J. Pharm. Sci.* **1995**, *84*, 697-706.
218. Tachibana, K.; Tachibana, S. *J. Pharm. Pharmacol.* **1991**, *43*, 270-271.
219. Tachibana, K. *Pharm. Res.* **1992**, *9*, 952-954.
220. Mitragotri, S.; Blankschtein, D.; Langer, R. *Science* **1995**, *269*, 850-853.
221. *Clin. Investigator News* **1995**, *3(10)*, 12-13.
222. Ritschel, W.A. *Meth. Find. Clin. Pharmacol.* **1991**, *13*, 205-220.
223. Schilling, R. J. Mitra, A. K. *Pharm. Res.* **1992**, *9*, 1003-1009.
224. Nishihata, T.; Rytting, J. H.; Kamada, A.; Higuchi, T. *Diabetes* **1981**, *30*, 1065-1067.

225. Saffran, M.; Kumar, G. S.; Savariar, C.; Burnham, J. C.; Williams, F.; Neckers, D. C. *Science* **1986**, *233*, 1081-1084.
226. Touitou, E.; Rubinstein, A. *Int. J. Pharm.* **1986**, *30*, 95-99.
227. Damge, C.; Michel, C.; Aprahamian, M.; Couvreur, P. *Diabetes* **1988**, *37*, 246-251.
228. Saffran, M.; Field, J. B.; Pena, J.; Jones, R. H.; Okuda, Y. *J. Endocrinol.* **1991**, *131*, 267-278.
229. Morishita, I.; Morishita, M.; Takayama, K.; Machida, Y.; Nagai, T. *Int. J. Pharm.* **1992**, *78*, 9-16.
230. Kraeling, M. E. K.; Ritschel, W. A. *Meth. Find. Clin. Pharmacol.* **1992**, *14*, 199-209.
231. Chang, C. L.; Gehrke, S. H.; Ritschel, W. A. *Meth. Find. Exp. Clin. Pharmacol.* **1994**, *16*, 271-278.
232. Scott-Moncrieff, J. C.; Shao, Z.; Mitra, A. K. *J. Pharm. Sci.* **1994**, *83*, 1465-1469.
233. Patel, H. M.; Ryman, B. E. *FEBS Letters* **1976**, *62*, 60-63.
234. Choudhari, K. B.; Labhasetwar, V.; Dorle, A. K. *J. Microencapsul.* **1994**, *11*, 319-325.
235. Weingarten, C.; Moufti, A.; Desjeux, J. F.; Luong, T. T.; Durand, G.; Devissaguet, J. P.; Puisieux, F. *Life Sci.* **1981**, *28*, 2747-2752.
236. Spangler, R. S. *Diabetes Care* **1990**, *13*, 911-922.
237. Mesiha, M. S.; el-Bitar, H. I. *J. Pharm. Pharmacol.* **1981**, *33*, 733-734.
238. Cho, Y. W.; Flynn, M. *Lancet* **1989**, *8678*, 1518-1519.
239. Hashizume, M.; Douen, T.; Murakami, M.; Yamamoto, A.; Takada, K.; Muranishi, S. *J. Pharm. Pharmacol.* **1992**, *44*, 555-559.
240. Mesiha, M. S. *Arch. Pharm. Chem. Sci. Ed.* **1981**, *9*, 137-142.
241. Mesiha, M.; Plakogiannis, F.; Vejosoth, S. *Int. J. Pharm.* **1994**, *111*, 213-216.
242. Singh, B. M.; Wise, P. H.; Marks, V. *Lancet* **1991**, *8762*, 308-309.
243. *Clin. Investigator News* **1996**, *4(6)*, 10.
244. Milstein, S.; Freire, E.; Sarubbi, D.; Leone-Bay, A. *The 3rd U.S.-Japan Symposium on Drug Delivery Systems*, **1995**, 15.
245. *Clin. Investigator News* **1996** *3(4)*, 9.
246. de Leede, L. G. J.; de Boer, A. G.; Roozen, C. P. J. M.; Breimer, D. D. *J. Pharm. Exp. Ther.* **1983**, *225*, 181-185.
247. Ichikawa, K.; Ohata, I.; Mitomi, M.; Kawamura, S.; Maeno, H.; Kawata, H. *J. Pharm. Pharmacol.* **1980**, *32*, 314-318.
248. Kim, S.; Kamada, A.; Higuchi, T.; Nishihata, T. *J. Pharm. Pharmacol.* **1983**, *35*, 100-103.
249. Yagi, T.; Hakui, N.; Yamasaki, Y.; Kawamori, R.; Shichiri, M.; Abe, H.; Kim, S.; Miyake, M.; Kamikawa, K.; Nishihata, T.; Kamada, A. *J. Pharm. Pharmacol.* **1983**, *35*, 177-178.
250. Nishihata, T.; Okamura, Y.; Inagaki, H.; Sudho, M.; Kamada, A.; Yagi, T.; Kawamori, R.; Shichiri, M. *Int. J. Pharm.* **1986**, *34*, 157-161.
251. Ritschel, W. A.; Ritschel, G. B.; Ristchel, B. E. C.; Lücker, P. W. *Meth. Find. Exp. Clin. Pharmacol.* **1988**, *10*, 645-656.
252. Ritschel, W. A.; Ritschel, G. B.; Sathyan, G. *Res. Comm. Chem. Path. Pharm.* **1988**, *62*, 103-112.
253. Ziv, E.; Lior, O.; Kidron, M. *Biochem. Pharmacol.* **1987**, *36*, 1035-1039.
254. van Hoogdalem, E. J.; Heijligers-Feijen, C. D.; Verhoef, J. C.; de Boer, A. G.; Breimer, D. D. *Pharm. Res.* **1990**, *7*, 180-183.
255. Nishihata, T.; Rytting, J. H.; Higuchi, T.; Caldwell, L. *J. Pharm. Pharmacol.* **1981**, *33*, 334-335.
256. Watanabe, Y.; Matsumoto, Y.; Hori, N.; Funato, H.; Matsumoto, M. *Chem. Pharm. Bull.* **1991**, *39*, 3007-3012.

257. Watanabe, Y.; Matsumoto, Y.; Seki, M.; Takase, M.; Matsumoto, M. *Chem. Pharm. Bull.* **1992**, *40*, 3042-3047.
258. Richardson J. L.; Illum, L.; Thomas, N. W. *Pharm. Res.* **1992**, *9*, 878-883.
259. Hosny, E.; el-Ahmady, O.; el-Shattawy, H.; Nabih, A.; el-Damacy, H.; Gamal-el-Deen, S.; el-Kabbany, N. *Arzneimittelforschung* **1994**, *44*, 611-613.
260. Yamasaki, Y.; Shichiri, M.; Kawamori, R.; Kikuchi, M.; Yagi, T.; Arai, S.; Tohdo, R.; Hakui, N.; Oji, N.; Abe, H. *Diabetes Care* **1981**, *4*, 454-458.
261. Okada, H.; Yamazaki, I.; Ogawa, Y.; Hirai, S.; Yashiki, T.; Mima, H. *J. Pharm. Sci.* **1982**, *71*, 1367-1371.
262. Lee, V. H. L. *Crit. Rev. Ther. Drug Carrier Systems* **1988**, *5*, 69-97.
263. Richardson, J. L.; Illum, L. *Adv. Drug Del. Rev.* **1992**, *8*, 341-366.
264. Okada, H.; Yashiki, T.; Mima, H. *J. Pharm. Sci.* **1983**, *72*, 173-176.
265. Okada, H.; Yamazaki, I.; Yashiki, T.; Mima, H. *J. Pharm. Sci.* **1983**, *72*, 75-78.
266. Golomb, G.; Avramoff, A.; Hoffman, A. *Pharm. Res.* **1993**, *10*, 828-833.
267. Ishida, M.; Machida, Y.; Nambu, N.; Nagai, T. *Chem. Pharm. Bull.* **1981**, *29*, 810-816.
268. Ritschel, W. A.; Ritschel, G. B.; Forusz, H.; Kraeling, M. *Res. Commun. Chem. Path. Pharm.* **1989**, *63*, 53-67.
269. Al-Achi, A.; Greenwood, R. *Res. Commun. Chem. Path. Pharmacol.* **1993**, *82*, 297-306.
270. Zhou, X. H.; Po, A. L. W. *Int. J. Pharm.* **1991**, *75*, 117-130.
271. Christie, C. D.; Hanzal, R. F. *J. Clin. Invest.* **1931**, *10*, 787-800.
272. Schoenwald, R. D. *Clin. Pharmacokinet.* **1990**, *18*, 255-269.
273. Nomura, M.; Kubota, M. A.; Kawamori, R.; Yamasaki, Y.; Kamada, T.; Abe, H. *J. Pharm. Pharmacol.* **1994**, *46*, 768-770.
274. Hayakawa, E.; Chien, D. S.; Inagaki, K.; Yamamoto, A.; Wang, W.; Lee, V. H. L. *Pharm. Res.* **1992**, *9*, 769-775.
275. Sasaki, H.; Tei, C.; Yamamure, K.; Nishida, K.; Nakamura, J. *J. Pharm. Pharmacol.* **1994**, *46*, 871-875.
276. Pillion, D. J.; Atchison, J. A.; Stott, J.; McCracken, D.; Gargiulo, C.; Meezan, E. *J. Ocular Pharm.* **1994**, *10*, 461-470.
277. Morgan, R. V. *J. Ocular Pharm.* **1995**, *11*, 565-573.
278. Pillion, D. J.; Recchia, J.; Wang, P.; Marciani, D. J.; Kesil, C. R. *J. Pharm. Sci.* **1995**, *84*, 1276-1279.
279. Sasaki, H.; Nagano, T.; Yamamura, K.; Nishida, K.; Nakamura, J. *J. Pharm. Pharmacol.* **1995**, *47*, 703-707.
280. Sasaki, H.; Tei, C.; Nishida, K.; Nakamura, J. *Biol. Pharm. Bull.* **1995**, *18*, 169-171.
281. Volkin, D.; Klibanov, A. M. In *Protein Function: A Practical Approach*; Creighton, T. E., Ed; IRL Press: Oxford, 1990, pp 1-24.
282. Brange, J. *Stability of Insulin. Studies on the Physical and Chemical Stability of Insulin in Pharmaceutical Formulation*; Kluwer: Boston, 1994.
283. Dudley, H. W. *Biochem. J.* **1923**, *17*, 376-390.
284. Jensen, H.; Evans, E. A., Jr.; Penington, W. D.; Schock, E. D. *J. Biol. Chem.* **1936**, *114*, 199-208.
285. du Vigneaud, V.; Fitch, A.; Pekaek, E.; Lockwood, W. W. *J. Biol. Chem.* **1931**, *94*, 233-242.
286. Wintersteiner, O. *J. Biol. Chem.* **1933**, *102*, 473-488.
287. Krough, A.; Hemmingsen, A. A. *Biochem. J.* **1928**, *2*, 1231-1238.
288. du Vigneaud, V.; Geiling, E. M. K.; Eddy, C. A. *J. Pharmacol. Exp. Ther.* **1928**, *33*, 497-509.
289. Gerlough, T. D.; Bates R. W. *J. Pharmacol. Exp. Ther.* **1932**, *45*, 19-51.
290. Waugh, D. F. A. *J. Am. Chem. Soc.* **1946**, *68*, 247-250.

291. Waugh, D. F.; Wilhelmson, D. F.; Commerford, S. L.; Sackler, M. L. *J. Am. Chem. Soc.* **1953**, *75*, 2592-2600.
292. Waugh, D. F. *Adv. Protein Chem.* **1954**, *9*, 325-437.
293. Jeffrey, P. D. *Biochemistry* **1974**, *13*, 4441-4447.
294. Milthorp, B. K.; Nichol, L. W.; Jeffrey, P. D. *Biochim. Biophys. Acta*, **1977**, *495*, 195-202.
295. Thurow, H.; Geisen, K. *Diabetologia* **1984**, *27*, 212-218.
296. Dathe, M.; Gast, K.; Zirwer, D.; Welfe, H.; Mehlis, B. *Int. J. Peptide Protein Res.* **1990**, *36*, 344-349.
297. Sluzky, V.; Tamada, J. A.; Klibanov, A. M.; Langer, R. *Proc. Natl. Acad. Sci. USA* **1991**, *88*, 9377-9381.
298. Albisser, A. M.; Williamson, J. R.; Lougheed, W. D. In *Hormone Drugs*; Gueriguian, J. L.; Bansome, E. D.; Outschoorn, A. S., Eds.; United States Pharmacopeial Convention: Rockville, Maryland, 1982, pp 84-95.
299. Lougheed, W. D.; Woulfe-Flanagan, H.; Clement, J. R.; Albisser A. M. *Diabetologia* **1980**, *19*, 1-9.
300. Pongor, S.; Brownlee, M.; Cerami, A. *Diabetes*, **1983**, *32*, 1087-1091.
301. Thurow, H. In *Insulin. Chemistry, Structure and Function of Insulin and Related Hormones*; Brandenburg, D.; Wollmer, A., Eds.; Gruyter: Berlin, 1980, pp 215-221.
302. Blackshear, P. J.; Rohde, T. D.; Palmer, J. L.; Wigness, B. G.; Rupp, W. M.; Buchwald, H. *Diabetes Care* **1983**, *6*, 387-392.
303. Sluzky, V.; Klibanov, A. M.; Langer, R. *Biotechnol. Bioeng.* **1992**, *40*, 895-903.
304. Bringer, J.; Heldt, A.; Grodsky, G. M. *Diabetes* **1981**, *30*, 353-359.
305. Sato, T.; Ebert, C. D.; Kim, S. W. *J. Pharm. Sci.* **1983**, *72*, 228-232.
306. Lougheed, W. D.; Albisser, A. M.; Martindale, H. M.; Chow, J. C.; Clement, J. R. *Diabetes*, **1983**, *32*, 424-432.
307. Quinn, R.; Andrade, J. D. *J. Pharm. Sci.* **1983**, *72*, 1472-1473.
308. Chawla, A. S.; Hinberg I.; Blais, P.; Johnson, D. *Diabetes* **1985**, *34*, 420-424.
309. Grau, U.; Geisen, K.; Jährling, P. *Diab. Nutr. Metab.* **1989**, *2*, 43-52.
310. James, D. E.; Jenkins, A. B.; Kraegen, E. W.; Chisholm, D. J. *Diabetologia* **1981**, *21*, 554-557.
311. Mecklenburg, R. S.; Guinn, T. S. *Diabetes Care* **1985**, *8*, 367-370.
312. Brange, J.; Havelund, S.; Hommel, E.; Sørensen, E.; Kühl, C. *Diabet. Med.* **1986**, *3*, 532-536.
313. Ratner, R. E.; Steiner, M. L. *Diabetes Care* **1987**, *10*, 787-788.
314. Eichner, H. L.; Selam, J.-L.; Woertz, L. L.; Cornblath, M.; Charle, M. A. *Diab. Nutr. Metab.* **1988**, *1*, 283-287.
315. Clark, S.; Stephenson, R. C.; Lowenson, J. D. In *Stability of Protein Pharmaceuticals. Part A. Chemical and Physical Pathways of Protein Degradation*; Ahern, T. J.; Manning, M. C., Eds.; Plenum Press: New York, 1992, pp 1-29.
316. Darrington, R. T.; Anderson, B. D. *Pharm. Res.* **1994**, *11*, 784-793.
317. Darrington, R. T.; Anderson, B. D. *Pharm. Res.* **1995**, *12*, 1077-1084.
318. Darrington, R. T.; Anderson, B. D. *J. Pharm. Sci.* **1995**, *84*, 275-282.
319. Brange, J.; Langkjær, L.; Havelund, S.; Vølund, A. *Pharm. Res.* **1992**, *9*, 715-726.
320. Brange, J.; Langkjær, L.; Havelund, S.; Sørensem, E. *Diabetes Res. Clin. Pract.* **1985**, *1* [*Suppl.*], 67.
321. Brange, J; Havelund, S.; Hougaard, P. *Pharm. Res.* **1992**, *9*, 727-734.
322. Robbins, D. C.; Cooper, S. M.; Fineberg, S. E.; Mead, P. M. *J. Clin. Invest.* **1986**, *77*, 717-723.
323. Fisher, B. V.; Porter, P. B. *J. Pharm. Pharmacol.* **1981**, *33*, 203-206.

324. Strickley, R. G.; Anderson, B. D. *Pharm. Res.* **1996**, 13, 1142-1153.
325. Strickley, R. G.; Anderson, B. D. *J. Pharm. Sci.* **1997**, in press.
326. Costantino, H. R.; Langer, R.; Klibanov, A. M. *Pharm. Res.* **1994**, *11*, 21-29.
327. Ahlneck, C.; Zografi, G. *Int. J. Pharm.* **1990**, *62*, 87-95.
328. Franks, F.; Hatley, R. H. M.; Mathias, S. F. *Biopharm.* **1991**, *4(9)*, 38-55.
329. Liu, W. R.; Langer, R.; Klibanov, A. M. *Biotechnol. Bioeng.* **1991**, *37*, 177-184.
330. Costantino, H. R.; Griebenow, K.; Mishra, P.; Langer, R.; Klibanov, A. M. *Biochim. Biophys. Acta* **1996**, *1253*, 69-74.
331. Prestrelski, S. J; Tedeschi, N.; Arakawa, T.; Carpenter, J. F. *Biophys. J.* **1993**, *65*, 661-671.
332. Prestreslki, S. J.; Arakawa, T.; Carpenter, J. F. *Arch. Biochem. Biophys.* **1993**, *303*, 465-473.
333. Hageman, M. J.; Bauer, J. M; Possert, P. L.; Darrington, R. T. *J. Agric. Food Chem.* **1992**, *40*, 348-355.
334. Schwartz, H. M.; Lea, C. H. *Biochem. J.* **1952**, *50*, 713-716.
335. Uchida, T.; Yagi, A.; Oda, Y.; Nakada, Y.; Goto, S. *Chem. Pharm. Bull.* **1996**, *44*, 235-236.
336. Schwendeman, S. P.; Costantino, H. R.; Gupta, R. K.; Langer, R. In *Controlled Drug Delivery. The Next Generation*; Park, K., Ed.; American Chemical Society: Washington, 1997, in press.

Chapter 3

Stability of the Dipeptide Aspartame in Solids and Solutions

Leonard N. Bell

Nutrition and Food Science, 328 Spidle Hall, Auburn University, AL 36849

Aspartame (APM) is a labile dipeptide whose stability is influenced by numerous factors, a discussion of which illustrates the challenges facing product development scientists. The activation energies for APM degradation decrease as either pH or moisture content increases. APM degradation is buffer-catalyzed and occurs faster in phosphate buffer than citrate buffer. Buffer concentration also influences the degradation of APM in low moisture solids. APM degradation rates in solid systems depend upon a_w rather than upon mobility limitations associated with the state of the system. In high moisture semi-solid gels, the rate of APM degradation is minimal between pH 3 and 5. The pH-rate profiles for APM degradation in high and low moisture systems are quite different. Dehydration and partial rehydration change both the rates and mechanisms of APM degradation in reduced-moisture solids due to solid state pH values differing from those initially of the solutions. Various excipients also influence APM stability. The properties of the peptide and system as well as potential interactions between the two need to be identified to maximize peptide stability.

The vast production of peptides by genetic engineering has introduced numerous challenges for pharmaceutical scientists, from delivery to bioavailability to stability of the product. This paper discusses the factors which influence peptide stability in solids and solutions, focusing specifically on the dipeptide aspartame (α-L-aspartyl-L-phenylalanine-1-methyl ester). However, before addressing the issues of stability, the degradation mechanism and kinetics of aspartame should be understood.

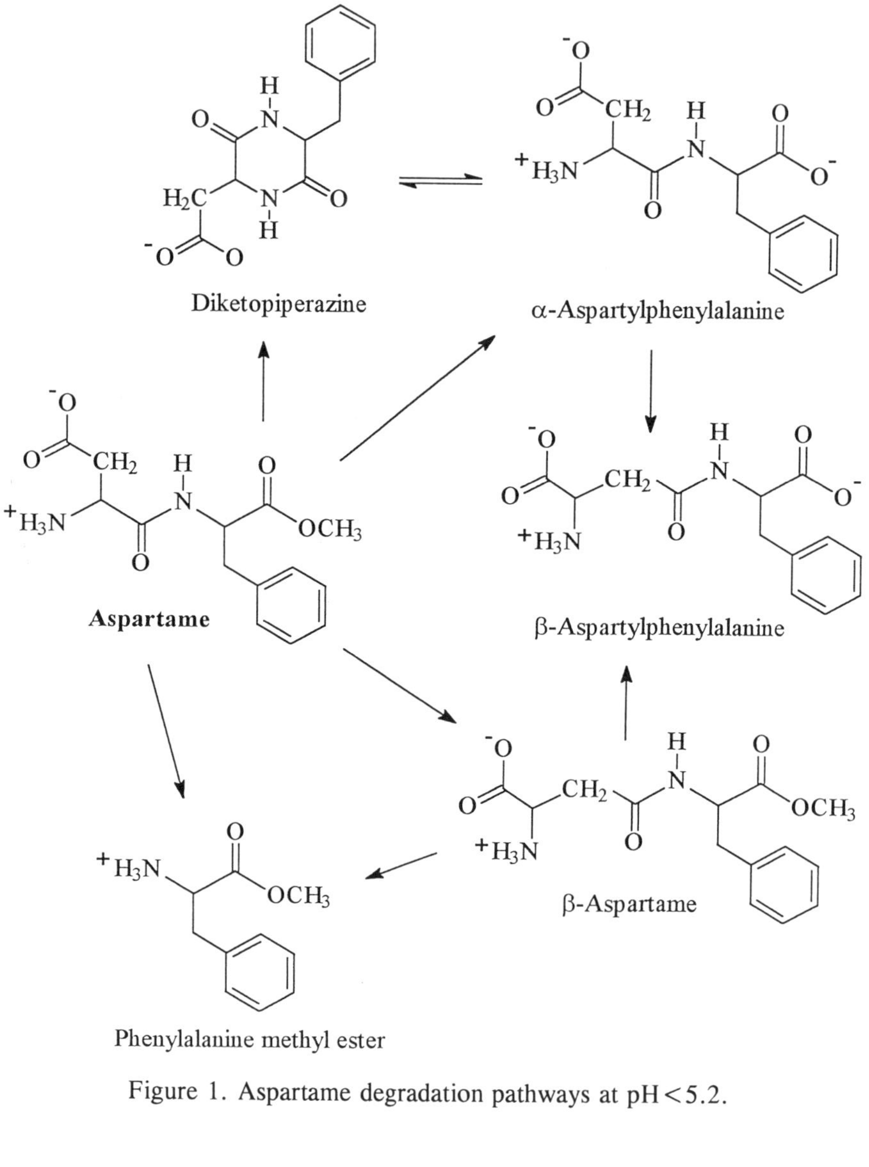

Figure 1. Aspartame degradation pathways at pH < 5.2.

Aspartame Degradation Kinetics

The degradation pathways of aspartame change as a function of pH. At pH values less than 5.2, aspartame forms α-aspartylphenylalanine (α-AP) via hydrolysis of the methyl ester group, diketopiperazine (DKP) via cyclization, phenylalanine methyl ester (PMe) via hydrolysis of the peptide bond, and β-aspartame (β-APM) via structural rearrangement through a cyclic intermediate (1-3). At these low pH values, β-aspartylphenylalanine (β-AP) is also produced from ester hydrolysis of β-aspartame and the rearrangement of α-AP (2). The aspartame degradation pathways at pH < 5.2 are shown in Figure 1. At pH > 5.2, aspartame degrades into DKP and α-AP without the formation of the β-isomers (1-3).

Based on the degradation products formed from aspartame at pH 7, the rate expression for the loss of aspartame can be written as

$$-d[APM]/dt = k_{DKP}[APM] + k_{\alpha\text{-}AP}[APM] \tag{1}$$

where k_{DKP} is the net rate constant for diketopiperazine formation, $k_{\alpha\text{-}AP}$ is the net rate constant for ester hydrolysis and consequent α-aspartylphenylalanine formation, and [APM] is the concentration of aspartame. For the acid-base catalyzed degradation of aspartame, each net rate constant, k_x, can be expressed in general terms as follows:

$$k_x = k_o + k_H[H^+] + k_{OH}[OH^-] + k_{BH}[BH] + k_B[B^-] \tag{2}$$

where k_o is the rate constant for the uncatalyzed reaction, k_H is the rate constant for catalysis by hydronium ions, k_{OH} is the rate constant for catalysis by hydroxyl ions, k_{BH} is the rate constant for buffer-mediated acid catalysis, and k_B is the rate constant for buffer-mediated base catalysis (4). In buffered solutions, the concentrations of the buffering components as well as the pH remain unchanged during the course of the reaction, which allows for the combination of rate constants to yield the pseudo first order expression

$$-d[APM]/dt = k_{obs}[APM] \tag{3}$$

where k_{obs} is the observed first order rate constant. The pseudo first order degradation model for aspartame in solutions and solids has been verified experimentally (1,3,5-11). Based on the above studies, the factors which affect aspartame stability in solutions and solids can be addressed.

Temperature

The effect of temperature on aspartame degradation has been evaluated in solutions and solid systems (1,6-9). Higher temperatures enhance peptide degradation rates, as expressed by the Arrhenius equation. The sensitivity of the reaction to temperature is given by the activation energy.

In solution, the activation energy for aspartame degradation decreased significantly with increasing pH (8-9). For example, Bell and Labuza found the activation energy changed from 19.7±1.4 kcal/mol at pH 3 to 15.2±0.8 kcal/mol at pH 7 in 0.1 M phosphate buffer (8). As mentioned previously, the pathways by which aspartame degrades change as a function of pH. Different pathways would each have different activation energies and the contribution from each pathway would differ with pH. Thus, an influence of pH on the activation energy for aspartame degradation would be expected.

In solid systems, the activation energies for aspartame degradation decreased with increasing moisture content (8). For example at pH 5, the activation energies for aspartame degradation decreased from 27.1±1.1 kcal/mol at 6% moisture to 24.5±1.3 kcal/mol at 14% moisture to 17.9±1.5 kcal/mol in solution (8). The decrease in activation energy with increasing moisture content has been explained using principles of enthalpy-entropy compensation (12-13). Assuming the degradation of aspartame requires the formation of an activated complex, then as moisture content increases, the entropy (ΔS) increases while the free energy (ΔG) decreases, resulting in a decrease in enthalpy (ΔH) and thus a lower activation energy (13). A detailed discussion of enthalpy-entropy compensation as applied to aspartame degradation can be found elsewhere (13).

The higher activation energy for reactions in low moisture solids can have definite implications with respect to product stability. If a product at 6% moisture (E_a = 27 kcal/mol) and as a solution (E_a = 18 kcal/mol) both have shelf lives of 12 months at 22°C, the low moisture system would have a longer shelf life than the high moisture product at refrigerated temperatures. However, under product abuse at 32°C, the low moisture system will lose shelf life prior to the high moisture system (2.7 months compared to 4.4 months). Low moisture products are usually more stable than their high moisture counterparts, which would not be the case for this product at 32°C.

Buffer Type and Concentration

In addition to temperature, the buffer type and concentration also influence the stability of aspartame in solutions as shown in Figure 2. Increasing the buffer concentration increased the rate of aspartame degradation (9,11). The buffer concentration logically influences the degradation rate of acid-base catalyzed reactions by increasing each net rate constant, k_x, as shown by equation 2. At pH 7 and phosphate buffer concentrations of 0.01 and 0.1 M, catalysis by the buffer components was responsible for approximately 87.3 and 98.5% of the degradation rate constant, respectively (11). Similar trends were also observed at pH 3 (11). Numerous other pharmaceuticals (e.g. ampicillin, chlorampenicol) are also sensitive to buffer concentration (14).

In solutions, phosphate buffer significantly enhanced the degradation of aspartame as compared to citrate buffer (11). The Brønsted equation,

$$\log(k_{obs}) = \log(C) + (z)\log(K_d) \qquad (4)$$

relates the rate constant (k_{obs}) of an acid-base catalyzed reaction, such as aspartame degradation, to the buffer ionization constant (K_d); C and z in equation 4 are constants (*4*). Figure 3 shows the Brønsted plot for aspartame degradation in 0.1 M buffer solutions at pH 7 and 25°C. Except for aspartame degradation in phosphate buffer, the log of the degradation rate constant changed linearly with the log of the buffer ionization constant (*11*). The degradation in phosphate buffer is faster than expected based on the buffer ionization constant alone, indicating that a factor other than the buffer ionization constant is enhancing the reaction. The effect of buffer type on aspartame degradation was explained by the ability of the small phosphate anion to simultaneously accept and donate the protons (i.e., bifunctional catalytic activity) required for the conversion of aspartame into diketopiperazine (*11*). The exchange of protons between citrate anions and aspartame is believed to occur less rapidly due to partial steric inhibition (*11*).

The effect of buffer on aspartame stability is not limited to solutions. Figure 4 shows the effect of the initial phosphate buffer concentration (i.e., prior to lyophilization and rehydration) on aspartame degradation in a solid system at pH 5 containing 6% moisture. The degradation rate in the low moisture system increased as buffer concentration increased (*8,15*). Products containing components susceptible to catalysis by buffer salts should be formulated at the lowest buffer concentration possible to minimize degradation rates and enhance shelf life.

Water Activity and Glass Transition

Water has many potential effects on the degradation of peptides. The influence of moisture on the folding and stability of polypeptides has been documented (*16-18*). This discussion will focus on the chemical potential of water (i.e., water activity) and the plasticization effect of water on amorphous solids. Water activity has been shown to correlate with reaction rates (*19*). In addition, the plasticization ability of water may convert amorphous glassy systems into amorphous rubbery systems upon moisture sorption, leading to enhanced mobility and potentially chemical reactivity as well (*20*).

Figure 5 shows the effect of water activity on the degradation of aspartame at pH 5 and 30°C. A 0.1 increase in water activity decreases the shelf life by approximately 40 to 50% (*10,13,21*). Moisture transfer into pharmaceutical solids during storage, which increases the water activity of the system, can enhance degradation rates (*14*).

The glass transition may influence chemical reactions if the mechanism has a mobility-dependent step. As mentioned, moisture sorption can convert a glassy matrix into a rubbery matrix of increased mobility, which may result in increased reaction rates. Using polyvinylpyrrolidone of different molecular weights, the influence of the glass transition on aspartame degradation was evaluated at constant temperature (*10*). Figure 6 shows that aspartame degradation at pH 7 is not influenced by the glass transition, but rather by the water activity. If mobility was rate-limiting, a dramatic increase in the rate constant would have occurred at the glass transition temperature ($T-T_g=0$). However, rate constants at constant water activity and different glass transition temperatures were similar whereas the rate

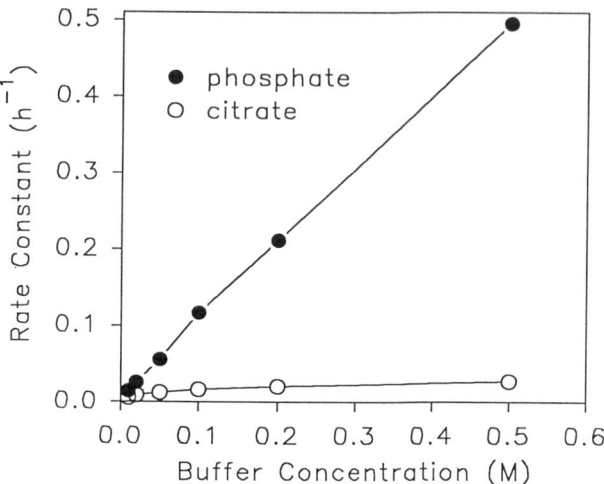

Figure 2. Effect of buffer type and concentration on the rate constant of aspartame degradation in solution at pH 7 and 25°C (Adapted from ref. 11).

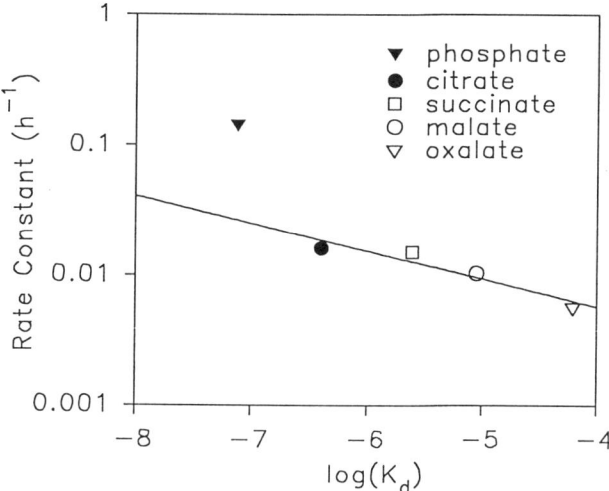

Figure 3. Brønsted plot of aspartame degradation in 0.1 M buffer at pH 7 and 25°C (Reproduced with permission from ref. 11. Copyright 1995 ACS).

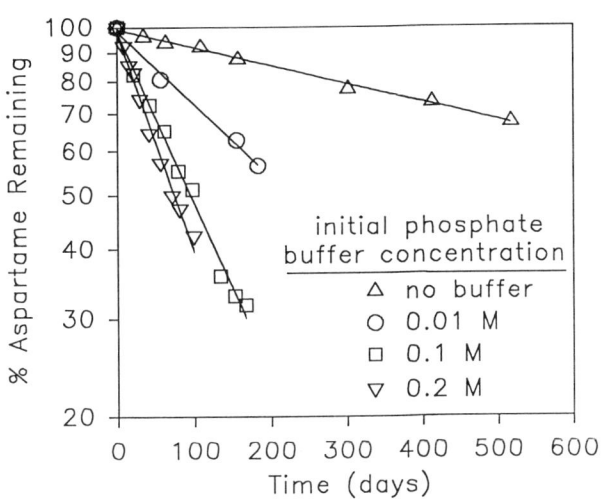

Figure 4. Aspartame degradation in a 6% moisture solid system at pH 5 and 30°C as influenced by initial phosphate buffer concentration (Adapted from ref. 8 and 15).

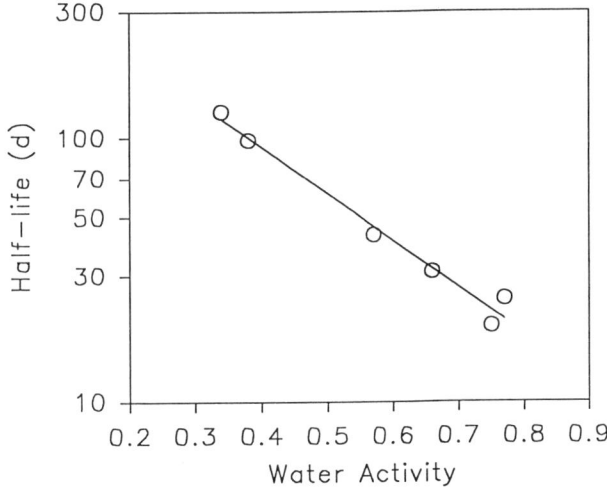

Figure 5. Relationship between aspartame half-life at 30°C and water activity in a lyophilized solid containing 0.1 M phosphate buffer at pH 5 prior to lyophilization (Adapted from ref. 8 and 15).

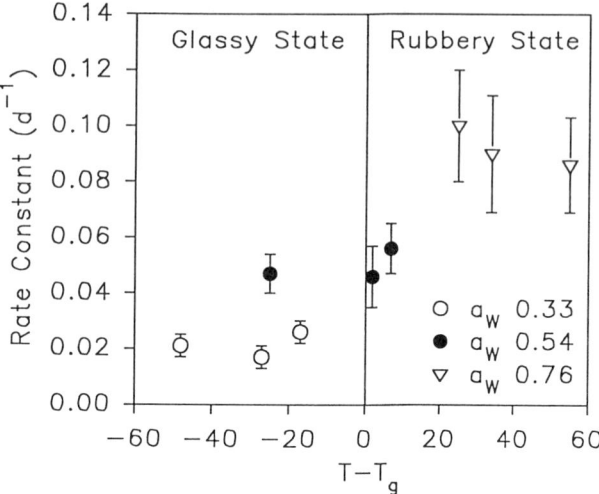

Figure 6. Aspartame degradation rate constants as influenced by the distance from the glass transition temperature in solids containing 0.1 M phosphate buffer at pH 7 and 25°C (Reproduced with permission from ref. 10. Copyright 1994 ACS).

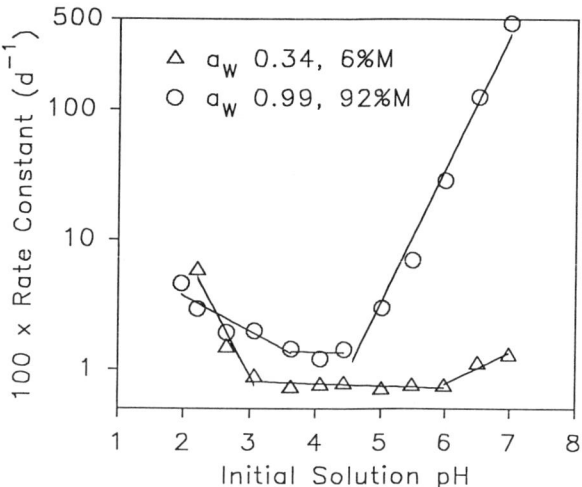

Figure 7. pH-rate profile for aspartame degradation in 0.1 M phosphate buffer at 30°C (Reproduced with permission from ref. 3. Copyright 1991 *Cryo-Letters*).

constants at different water activities and similar glass transition temperatures were significantly different (10). The amount of free volume in the solid system is sufficient for the localized conformational flexibility necessary for aspartame to cyclize into DKP. Long range molecular mobility (e.g., diffusion) required for other types of reactions would probably be more dependent on whether the system was in a glassy or rubbery state.

pH

The impact of pH on aspartame degradation has also been examined in liquid and solid systems. Aspartame is most stable between pH 3.5 and 5, with increased acid hydrolysis at lower pH values and increased base catalysis at higher pH values (1,3,5-6,8-9,15,22). Figure 7 shows the U-shaped pH-rate profiles for aspartame degradation in a high moisture semi-solid gel (92% moisture) and a low moisture solid (6% moisture). In the low moisture system, rate constants are generally lower, as expected based on the previous water activity discussion. However, the profile for the low moisture system also has a broader minima plateau than that of the high moisture system and a lower slope on the base-catalyzed portion of the profile (3,15).

The difference between the two profiles can be explained by evaluating the pH of the low moisture system. The degradation products formed in the high moisture gel and at 6% moisture are listed as a function of the initial solution pH in Table I. In the high moisture gel, the β-isomers are formed at pH values of 5 and lower while at 6% moisture the β-isomers are also formed at higher pH values (e.g., initial solution pH of 6.5) where these products are not expected to be formed. One explanation for the appearance of β-isomers at the higher pH values in the 6% moisture system could be that the pH of this system decreased upon lyophilization

Table I. Aspartame Degradation Products Formed at 30°C

Initial Solution pH	Semi-solid Gel (92% moisture)	Solid (6% moisture)
2.6	DKP, α-AP, β-AP, β-APM, PMe	DKP, α-AP, β-APM
3.0	DKP, α-AP, β-AP, β-APM, PMe	DKP, α-AP, β-AP, β-APM
4.4	DKP, α-AP, β-AP, β-APM	DKP, α-AP, β-AP, β-APM
5.0	DKP, α-AP, β-AP, β-APM	DKP, α-AP, β-AP, β-APM
5.5	DKP, α-AP	DKP, α-AP, β-AP, β-APM
6.0	DKP, α-AP	DKP, α-AP, β-AP, β-APM
6.5	DKP, α-AP	DKP, α-AP, β-APM

Adapted from ref. 3 and 15.

and rehydration (*3,15*). Similarly, Table II shows the DKP:α-AP concentration ratio as a function of the initial solution pH. These ratios from the 6% moisture system at higher pH values are similar to the ratios from the high moisture system at lower pH values, which suggests that the pH decreased by about 2 units in the low moisture system (*3,15*). This lowering of pH may be attributable to selective precipitation of buffer salts (i.e., solubility limitations) and the increased concentration of protons in the limited aqueous phase of the low moisture solid (*3,23-25*).

Table II. DKP:α-AP Ratio from Aspartame Degradation at 30°C

Initial Solution pH	Semi-solid Gel (92% Moisture)	Solid (6% Moisture)
2.0	0.2	N.D.[a]
2.6	0.5	0.2
3.0	1.0	0.3
4.4	2.0	0.3
5.0	3.0	0.6
5.5	5.0	0.6
6.0	22.0	0.7
6.5	75.0	3.0

[a]N.D. = not determined; Adapted from ref. 3 and 15.

Excipient Interactions

Most of the studies examining interactions between aspartame and other substances have focused on carbonyl compounds. In the presence of glucose, aspartame has been shown to participate in the Maillard reaction at elevated temperatures (*26-27*). It has also been suggested that other carbonyls (e.g., vanillin) react similarly with aspartame in solution (*28*). In a 0.1 M phosphate buffer solution at pH 5 and 30°C, 0.3 mM aspartame degraded faster in the presence of 0.8 mM glucose than in a non-glucose containing solution (Bell, L. N., University of Minnesota, unpublished data). In reduced-moisture solid systems, the addition of glucose alone did not enhance the loss of aspartame (*21*). However, reduced-moisture solid systems containing both glucose and emulsified oil did display faster aspartame degradation rates as compared to similar systems of aspartame alone or aspartame with glucose (*21*). While the mechanism by which oil enhanced the degradation of aspartame is unknown, the oil may have provided an additional reaction medium or the oil and water may have formed a reactive interface (*21*).

A limited number of studies have examined the interaction of aspartame with substances other than buffer salts or carbonyls. At 30°C, aspartame in 0.1 M phosphate buffer solutions at pH 3, 5, and 7 (8) had degradation rate constants similar to the same solution formed as an agar/microcrystalline cellulose gel (3), which suggests these carbohydrate polymers are inert with respect to aspartame stability. The addition of calcium caseinate to 0.01 M phosphate buffer at pH 7 and 80°C decreased the aspartame degradation rate; the mechanism for this protective effect is unknown (29). Aspartame in solution is also stabilized by complexation with β-cyclodextrin (30). It is clearly worthwhile to understand the effect excipients have on peptide stability during product formulation.

Summary

The formulation of peptide products is challenging and requires a thorough understanding of the factors which influence peptide stability as well as how stability differs in solids and liquids. Aspartame stability data can provide some useful information regarding problems that may arise during formulation. Due to different activation energies at different pH values and moisture contents, the activation energy for the reaction in the specific formulation should be used to accurately predict product shelf life. Buffer salts catalyze the degradation of aspartame; the potential effects of buffer type and concentration need to be determined to minimize degradation rates and enhance shelf life. Water activity correlates with aspartame stability. However, depending on the mechanism of degradation and the extent of movement required for degradation, glass transition may influence the degradation of larger peptides. Understanding whether water activity or glass transition influences chemical reactions is necessary for optimizing solid state peptide stability. Changes in pH upon lyophilization not only change the rate of aspartame degradation but the mechanism by which it degrades. Peptide-based products should therefore be formulated such that the final pH is optimized rather than the initial solution pH of the system. Carbonyl compounds, due to the Maillard reaction, can enhance peptide degradation, depending upon the moisture content of the system. However, caseinates and cyclodextrins can reduce APM degradation. Thus, the effect of excipients on peptide stability must also be evaluated. By keeping these factors in mind during formulation, the pharmaceutical scientist can develop peptide products of optimum stability.

Acknowledgments

This work is supported by project ALA-13-005 of the Alabama Agricultural Experiment Station and is published as AAES Journal No. 10-965200.

Literature Cited

1. Prudel, M.; Davidkova, E.; Davidek, J.; Kminek, M. *J. Food Sci.* **1986**, *51*, 1393-1415.
2. Stamp, J.A.; Labuza, T.P. *Food Add. Contam.* **1989**, *6*, 397-414.

3. Bell, L.N.; Labuza, T.P. *Cryo-Lett.* **1991**, *12*, 235-244.
4. Alberty, R.A. *Physical Chemistry*; Wiley: New York, NY, 1987; pp 760-762.
5. Prudel, M.; Davidkova, E. *Die Nahrung* **1981**, *25*, 193-199.
6. Ozol, T. *Acta Pharm. Turc.* **1986**, *28*, 125-130.
7. Araman, A.; Temiz, D. *Acta Pharm. Turc.* **1988**, *30*, 28-32.
8. Bell, L.N.; Labuza, T.P. *J. Food Sci.* **1991**, *56*, 17-20.
9. Tsoubeli, M.N.; Labuza, T.P. *J. Food Sci.* **1991**, *56*, 1671-1675.
10. Bell, L.N.; Hageman, M.J. *J. Agric. Food Chem.* **1994**, *42*, 2398-2401.
11. Bell, L.N.; Wetzel, C.R. *J. Agric. Food Chem.* **1995**, *43*, 2608-2612.
12. Labuza, T.P. *Food Technol.* **1980**, *34(2)*, 67-77.
13. Bell, L.N.; Labuza, T.P. *J. Food Eng.* **1994**, *22*, 291-312.
14. Conners, K.A.; Amidon, G.L.; Stella, V.J. *Chemical Stability of Pharmaceuticals, 2nd ed.*; Wiley: New York, NY, 1986.
15. Bell, L.N. *Investigations regarding the definition and meaning of pH in reduced-moisture model food systems*; UMI: Ann Arbor, MI; 1992, pp 109-158.
16. Hageman, M.J. *Drug Dev. Ind. Pharm.* **1988**, *14*, 2047-2070.
17. Hageman, M.J. In *Stability of Protein Pharmaceuticals, Part A: Chemical and Physical Pathways of Protein Degradation*; Ahern, T.J.; Manning, M.C., Ed.; Plenum: New York, NY, 1992; pp 273-309.
18. Zaks, A. In *Stability of Protein Pharmaceuticals, Part A: Chemical and Physical Pathways of Protein Degradation*; Ahern, T.J.; Manning, M.C., Ed.; Plenum: New York, NY, 1992; pp. 249-271.
19. Labuza, T.P. *Food Technol.* **1980**, *34(4)*, 36-41, 59.
20. Roos, Y.H. *Phase Transitions in Foods*; Academic Press: San Diego, 1995.
21. Bell, L.N.; Labuza, T.P. In *Water Relationships in Foods*; Levine, H.; Slade, L., Ed.; Plenum: New York, NY, 1991; pp 337-349.
22. Skwierczynski, R.D.; Conners, K.A. *Pharm. Res.* **1993**, *10*, 1174-1180.
23. van den Berg, L.; Rose, D. *Arch. Biochem. Biophys.* **1959**, *81*, 319-329.
24. van den Berg, L. *J. Dairy Sci.* **1961**, *44*, 26-31.
25. Bell, L.N.; Labuza, T.P. *J. Food Sci.* **1992**, *57*, 732-734.
26. Stamp, J.A.; Labuza, T.P. *J. Food Sci.* **1983**, *48*, 543-544, 547.
27. Huang, T.-C.; Soliman, A.A.; Rosen, R.T.; Ho, C.-T. *Food Chem.* **1987**, *24*, 187-196.
28. Tateo, F.; Triangeli, L.; Panna, E.; Berte, F.; Verderio, E. In *Frontiers of Flavor*; Charalambous, G., Ed.; Elsevier: Amsterdam, Netherlands, 1988; pp 217-231.
29. Tsoubeli, M.N.; Labuza, T.P. *J. Food Sci.* **1992**, *57*, 361-365.
30. Prankerd, R.J.; Stone, H.W.; Sloan, K.B.; Perrin, J.H. *Int. J. Pharm.* **1992**, *88*, 189-199.

Chapter 4

Mechanisms of Methionine Oxidation in Peptides

Christian Schöneich, Fang Zhao, Jian Yang[1], and Brian L. Miller

Department of Pharmaceutical Chemistry, School of Pharmacy,
University of Kansas, 2095 Constant Avenue, Lawrence, KS 66047

We review mechanistic details on the oxidation of methionine (Met) residues in representative model peptides. Peptides were subjected to oxidation by specific reactive oxygen species in order to obtain information on whether those species participate in more complex systems such as the metal-catalyzed oxidation of peptides, relevant to the pharmaceutical formulation of peptides and proteins as they are often contaminated with metals. Emphasis is placed on the peptide sequence determining which potentially catalytically active or inactive residues in the immediate vicinity of a Met residue may influence its oxidation pattern. Neighboring group effects play a significant role in product formation as they influence the chemistry of reactive intermediates formed at a given Met residue. For example, during the oxidation of Thr-Met such effects promote the side chain cleavage of the Thr residue following an initial attack of a hydroxyl radical at Met, and for Met-Met diastereoselectivity is observed with respect to the formation of azasulfonium derivatives.

A major problem associated with the manufacturing and formulation of peptides and proteins is the oxidative degradation of oxidation labile amino acids. Based solely on their redox potentials (*1*), we expect that the amino acids cysteine, tryptophan, tyrosine, histidine, and methionine will be most sensitive to oxidation, consistent with many observations on the *in vitro* and *in vivo* oxidation of these amino acid residues in proteins and peptides (*2–4*). However, it will strongly depend on the nature as well as the location of formation of an oxidizing species whether only the most sensitive amino acids or potentially also other amino acid residues will suffer

[1]Current address: Amgen, Inc., 1840 DeHavilland Drive, Thousand Oaks, CA 91320

oxidation in a given peptide or protein. It was shown that a rather nonselective strong oxidant such as the hydroxyl radical (HO•), produced by γ-irradiation of aqueous protein solutions, will attack proteins randomly and will oxidize amino acids based on their location within surface accessible domains of proteins (5). More selective oxidants such as peroxides will predominantly react with methionine and cysteine where surface exposure as well as the protonation state of cysteine will be important parameters. Under conditions of metal-catalyzed oxidation a peroxide may first react with a transition metal, bound to a metal-binding site of a protein, before the resulting reactive oxygen species will predominantly attack amino acids in the surrounding of the metal-binding site.

In addition, chain oxidations and radical transfer reactions within the peptide or protein will lead to even more complex oxidation patterns. For example, protein-bound peroxides (6) and in particular valine hydroperoxide (7) have been identified as important endpoints of hydroxyl radical-induced reactions. However, these hydroperoxides will only be true endpoints in the absence of catalytically active transition metals which can convert hydroperoxides into highly reactive alkoxyl radicals in a Fenton type reaction, initiating further protein damage through subsequent reactions of the alkoxyl radicals (8, 9). Consequently, an initial oxidative event at a protein locus can initiate protein chain oxidations (10) where oxidation products can be expected to appear remote from the side of the initial attack. Furthermore, several studies on electron transfer processes in proteins have now demonstrated that intramolecular electron transfer (or hydrogen transfer) may lead to the migration of an initial oxidation site (11). The final oxidation products of a given peptide or protein will, therefore, be the complex result of (i) the nature of the oxidizing species, (ii) the nature of the peptide or protein, determining sequence and higher order structure, and (iii) the mechanism of the reaction of the oxidizing species with the peptide or protein as it may depend on a specific peptide or protein sequence.

In pharmaceutical formulations it is generally difficult to characterize the nature of an oxidizing species and, consequently, the mechanism of its reaction with a peptide or protein. However, details about these mechanisms would be desirable in order to assist in the development of stable peptide and protein formulations. Some information about potential reactions can be obtained from the biochemical literature where a series of proteins has been investigated with respect to metal-catalyzed oxidation (3, 12–14), photooxidation (15–17), and (γ-radiolysis (6, 10, 18–20). In the chemical literature, we find more detailed investigations on chemical kinetics, neighboring group effects, and mechanisms of oxidation of small peptides and amino acids (or their derivatives) by various reactive oxygen species and free radicals (see below). Ideally, we would need to characterize the chemical mechanisms of oxidation by various reactive oxygen species or free radicals for a given peptide and then compare the kinetic features and product distributions for the oxidation of the same peptide within a pharmaceutical formulation. Whenever the product distributions obtained through both approaches match there will be a probability that the underlying reaction mechanisms occurring within a formulation have been identified. However, this approach will not work whenever several reactive oxygen species give the same product with a given amino acid. This will be the case, for example, for methionine and its major oxidation product, methionine sulfoxide, or for cysteine, and its products cystine or cysteine sulfonic acid. In such cases, we need more refined experiments in order to identify the contributing reactive intermediates. Several scavengers can be employed which will either selectively react with one reactive species (e.g. superoxide dismutase with superoxide) or at least will show different kinetics in their reactions with different reactive oxygen species. Whenever rate constants are available we can then predict the efficiency of a scavenger towards a reactive species and compare theoretical with experimental data. In the following, we shall discuss some selected examples for these approaches as they relate to the oxidation of methionine in small peptides. For this purpose we shall first introduce several important mechanistic aspects on the free radical chemistry of some model peptides before the discussion of some pharmaceutically relevant aspects of these considerations will follow. The investigated peptides have been selected on the basis of their model character and/or their presence in an oxidation sensitive domain of calmodulin which was identified to be highly oxidation labile both *in vitro* (*21–24*) and *in vivo* (*25*). They do as yet not represent pharmaceutical examples. However, as we will proceed from small model peptides to larger structures in future experiments, pharmaceutically relevant peptides will become a major focus of investigation, and the mechanisms obtained with small model peptides are likely to be an important basis for these studies.

Oxidation of Thr-Met: The migration of radical sites

Mechanistic studies with the hydroxyl radical. When the peptide Thr-Met was reacted with radiation chemically produced hydroxyl radicals (HO•) in anaerobic reaction mixtures at neutral pH, a major oxidation product was acetaldehyde, derived from the fragmentation of the Thr side chain (*26*). This reaction did not occur whenever the C-terminal Met residue was omitted, as in Thr-Leu, or when Thr was not the N-terminal amino acid, as in Gly-Thr-Met. Moreover, acetaldehyde was not a major reaction product when Thr-Met was subjected to one-electron oxidation by $SO_4^{\bullet-}$ (*26*) or by triplet carboxybenzophenone (*27*), indicating that it was not an initial one-electron transfer process, i.e. the formation of a sulfur radical cation, which led to the observed product formation.

Mechanistically, the reaction was rationalized by a fast formation of a hydroxysulfuranyl radical **1** upon reaction of the hydroxyl radical with the Met residue of Thr-Met (Scheme 1). The hydroxysulfuranyl radical subsequently underwent rapid formation of a three-electron-bonded intramolecular $[S\therefore N]^+$ complex **2** (reaction 1) which was characterized by the fast method of pulse radiolysis and had a half-life for further decomposition of $t_{1/2} = 320$ ns. An important equilibrium (reaction 2) of complex **2** is the open chain species containing a nitrogen-centered radical cation, a precursor for efficient heterolytic cleavage of the C_α–C_β bond of the Thr side chain leading to acetaldehyde and a proton (though some cleavage may directly occur from the cyclic complex **2**) (reaction 3).

This mechanism bears several interesting features that are important to consider in any study on peptide oxidation. First, from the published rate constants for the reaction of hydroxyl radicals with methionine, it was expected that a major fraction of hydroxyl radicals (>93%) reacted with the methionine moiety of the peptide. Nevertheless, the major final reaction product was a species containing intact Met but a fragmented Thr residue, indicated by the high yields of acetaldehyde (for [Thr-

Met] ≤ 5 × 10^{-4} M: 60% related to the initial yield of HO•). Such radical or damage transfer constitutes an important mechanistic pathway frequently observed during the oxidation of proteins.

Scheme 1

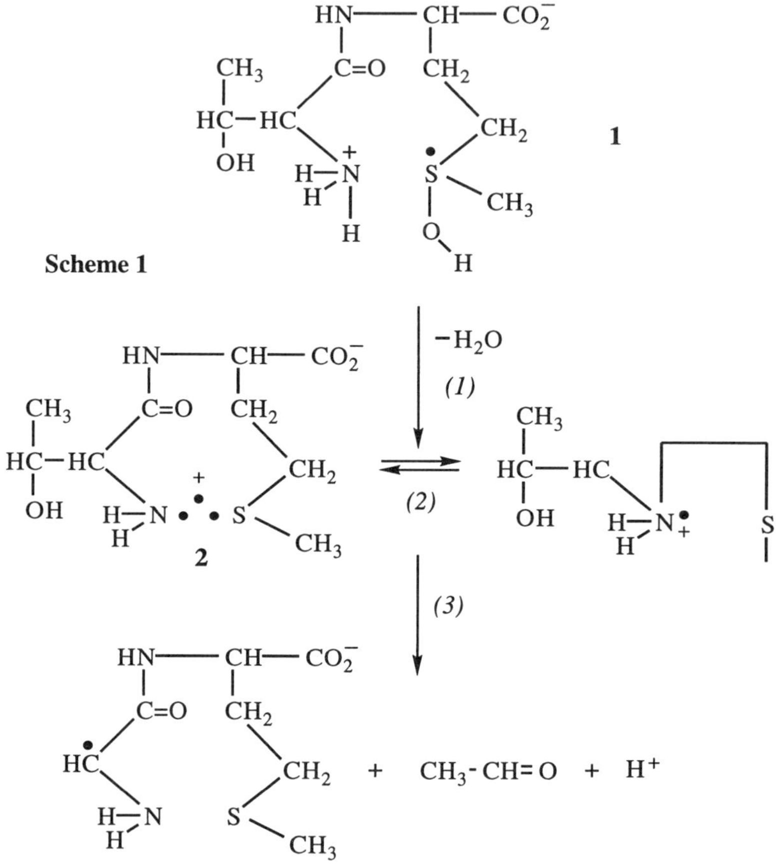

In the absence of any detailed mechanistic information, the lack of oxidized Met residues as major products may have led to the conclusion that Met was not a target during the oxidation. However, as demonstrated above this is not true. In fact, Met constituted a major point of attack for the reactive oxygen species, but radical transfer reactions subsequent to Met oxidation "repaired" the initial oxidation product (the sulfuranyl radical) and the final radical damage appeared at a different locus of the peptide. Besides for Thr-Met we could demonstrate a similar radical transfer reaction also for Ser-Met (26). In addition, comparable radical transfer mechanisms within some S-alkylglutathionine derivatives (28) and γ-Glu-Gly-Met-Gly (29) led to an efficient decarboxylation of the N-terminal γ-Glu residue, mediated by an initially formed hydroxysulfuranyl radical at the Met residue.

Returning to the example of Thr-Met, there were several other mechanistically interesting observations. First, in the presence of oxygen the yields of acetaldehyde were significantly reduced (by ca. 50%). This finding led to the hypothesis that hydroxysulfuranyl radicals might react competitively with molecular oxygen with the respective products entering reaction pathways different from acetaldehyde formation. Indeed, we could demonstrate that a model hydroxysulfuranyl radical does react with molecular oxygen with a rate constant on the order of 1.1×10^8 $M^{-1}s^{-1}$ (*30, 31*). Since free superoxide was not an immediate product of this reaction, this process likely involves the addition of oxygen to the hydroxysulfuranyl radical, resulting in a peroxyl type radical (reaction 4).

$$R_2S^{\bullet}-OH + O_2 \rightarrow HO-S(R_2)-OO^{\bullet} \qquad (4)$$

The structures of related peroxyl radicals have subsequently been characterized by low temperature ESR experiments (*32*). However, the rate constant for reaction 4 appears too slow in order to affect the formation of acetaldehyde in the Thr-Met system so that the effect of oxygen on the yields of acetaldehyde should have additional reasons which have yet to be characterized.

$$H_3\overset{+}{N}\text{-(Thr)-CONH}-\underset{\underset{\underset{H_3C}{\diagup}\overset{+}{S}\cdot\cdot S\diagdown_{CH_3}}{|}}{\overset{CO_2^-}{\underset{|}{CH}}} \qquad \overset{CO_2^-}{\underset{|}{HC}}-\text{NH-CO-(Thr)-}\overset{+}{N}H_3$$

3

Secondly, when the oxidation of Thr-Met by HO• was carried out with different concentrations of Thr-Met it was observed that increasing concentrations of Thr-Met led to decreasing yields of acetaldehyde but increasing yields of a one-electron oxidized sulfur species, an intermolecular sulfur radical cation complex **3** (this species was, again, identified by pulse radiolysis). The latter exists in equilibrium 5 with the monomeric radical cation **4** (>S•+) both of them being good one-electron oxidants themselves (*33*). However, we note that the radical complex **3** is significantly more stable than the monomer **4** with regard to irreversible decomposition pathways such as deprotonation (yielding α-(alkylthio)alkyl radicals). Thus, as

$$[>S\therefore S<]^+ \quad (3) \quad = \quad >S \quad + \quad >S^{\bullet+} \quad (4) \qquad (5)$$

already concluded from the experiments with $SO_4^{\bullet-}$ and triplet carboxybenzophenone, any one-electron oxidation of the Met residue of Thr-Met, yielding a sulfur radical cation, yields acetaldehyde less efficiently or not at all (depending on the initial oxidant). Although the sulfur atom in the hydroxysulfuranyl radical **1** formally contains a one-electron oxidized sulfur (compared to the reactant thioether), its chemistry is interestingly quite different compared to that of another one-electron oxidized sulfur species, the sulfur radical cation. The similarity of both may be better realized by considering that both the hydroxysulfuranyl radical **1** and the monomeric radical cation **4** are related to each other in an acid-base type equilibrium (reaction 6). Sulfur radical cations have theoretically been shown to exist in a hydrated form **4a** where the S–O bond dissociation energy amounts to ca. 16.8 kcal/mole (*34*). Formally, species **4a** has only to deprotonate in order to transform into **1**, although some electronic (and geometric) rearrangement may also be required, depending on the actual type of the hydroxysulfuranyl radical, i.e. σ, π, or σ* radical. Thus, for Thr-Met, a simple protonation-deprotonation equilibrium such as reaction 6 may eventually decide on the fate of the initially formed intermediate at the sulfur.

$$[>S^\bullet-OH_2]^+ \quad (4a) \quad = \quad H^+ \quad + \quad >S^\bullet-OH \ (1) \qquad (6)$$

Metal-catalyzed oxidation. With this mechanistic framework at hand it was now possible to investigate the oxidation of Thr-Met by hydrogen peroxide in the presence of transition metals. The transition metal/hydrogen peroxide system is of great importance for pharmaceutical formulations as they often contain metal impurities through buffer contamination and traces of peroxides from sterilization procedures. Moreover, peptides themselves are well suited ligands for many transition metals so that metal chelation by peptides may easily transfer transition metals from production lines to storage/formulation systems.

Table 1 contains a list of some key products obtained during the reaction of Thr-Met with HO•, hydrogen peroxide in the presence of ferrous iron, and hydrogen peroxide in the presence of ferrous EDTA (*35*). It becomes immediately apparent that the incubation of hydrogen peroxide with ferrous iron alone did not result in a product distribution which resembles the one obtained through free hydroxyl radicals. Thus, the classical Fenton system does not produce free hydroxyl radicals, at least at pH 7.5. Nevertheless we find a significant formation of Met sulfoxide. Moreover, the sulfoxide formation by the hydrogen peroxide/ferrous iron system is very rapid as the reaction was completed within ca. 2 min for 5.0×10^{-4} M hydrogen peroxide, 5.0×10^{-4} M ferrous iron, and 1.0×10^{-3} M Thr-Met. In contrast, the oxidation of 1.0×10^{-3} M Thr-Met by 5.0×10^{-4} M hydrogen peroxide alone did not

Table 1: Oxidation of Thr-Met by free hydroxyl radicals and Fenton systems

	HO•[a]	Fe^{II}/H_2O_2[b]	$[Fe^{II}(EDTA)]^{2-}/H_2O_2$[b]
-Thr-Met[c]	63	232±57	354 ± 57
Thr-Met(SO)	≤5.1	164±15	66 ± 12
CH_3CHO	36	6±3	65 ± 5
CO_2	13.2[d]	n.d.	n.d.

[a]Yields are from γ–radiolysis employing a dose of 100 Gly (calculated from reference 26).
[b]pH 7.5; 2×10^{-3} mol dm^{-3} carbonate buffer. [c]Refers to loss of Thr-Met. [d]Estimated from reference 36

SOURCE: Adapted from reference 35.

result in any significant formation of sulfoxide within 2 minutes, demonstrating that, although not generating hydroxyl radicals, nevertheless ferrous iron catalyzes the oxidation of Thr-Met by hydrogen peroxide. Thus, impurities of transition metals can catalyze peroxide mediated peptide oxidation via mechanisms quite different from the classical Fenton reaction. This does not only apply to small dipeptides as we have also observed extremely rapid oxidation of methionine-containing dodecapeptides (to methionine sulfoxide) by hydrogen peroxide/ferrous iron (unpublished results). Mechanistically, we propose that hydrogen peroxide merely binds to the transition metal, generating a complex that transfers oxygen to the Met residue. It seems reasonable to assume that these reactions are not only catalyzed by the metal *per se* but also further accelerated by the binding of the transition metal to the peptide, promoting an intramolecular reaction between the ferrous iron-hydrogen peroxide complex and the Met residue. In fact, some evidence for the latter was obtained when the presence of an additional His residue promoted the iron-catalyzed oxidation of nearby Met residues (*37, 38* (although these oxidations were carried out by a rather unspecific oxidizing system consisting of either dithiothreitol or ascorbate, ferric iron, and molecular oxygen). It is known that proteins which contain well-defined metal binding sites are prone to site-specific oxidation by reactive oxygen species generated within these metal-binding sites (*3, 12, 13*). However, in most cases these metal-binding sites require an intact secondary and tertiary structure of the protein. Thus, specific peptides comprising only one or the other metal-binding amino acid residue of a metal-binding site may not necessarily suffer site-specific oxidation like the intact metal-binding domain does. Such a case was recently observed for the metal-catalyzed oxidation of recombinant human growth hormone (*39*).

Table 1 shows that the presence of EDTA did not protect Thr-Met from oxidation by hydrogen peroxide/ferrous iron. However, it becomes apparent that significant yields of acetaldehyde are formed in this system. Thus, either hydroxyl radicals or hydroxyl radical-like species are formed in the [Fe^{II}(EDTA)]$^{2-}$/hydrogen peroxide system. Further evidence for the presence of hydroxyl radical-like species in the [Fe^{II}(EDTA)]$^{2-}$/hydrogen peroxide system was derived from a variation of the Thr-Met concentration. If hydroxyl radicals or alike species were formed in our systems we would expect higher relative yields of acetaldehyde at lower Thr-Met concentrations (as described above for free hydroxyl radicals). This was indeed observed, supporting the conclusion that at least a fraction of hydroxyl radicals or alike species are present in our [Fe^{II}(EDTA)]$^{2-}$/hydrogen peroxide system, responsible for acetaldehyde formation. On the other hand, significant yields of Met sulfoxide are also present after oxidation of Thr-Met by [Fe^{II}(EDTA)]$^{2-}$/hydrogen peroxide. Earlier, we had shown that the reaction of free hydroxyl radicals with aliphatic sulfides did not result in sulfoxide formation, particularly in the absence of oxygen (*40*). When we oxidized Thr-Met by the [Fe^{II}(EDTA)]$^{2-}$/hydrogen peroxide system in the additional presence of 2-propanol (a well-characterized hydroxyl radical scavenger), we could demonstrate that the obtained product yields could not be rationalized by competition kinetics based on a competitive reaction of hydroxyl radicals with either Thr-Met or 2-propanol. Thus, the yields of Thr-Met sulfoxide have to be ascribed to a reactive oxygen species being of different nature than a hydroxyl radical.

These results have at least two major implications for pharmaceutical formulations. First, the presence of a metal chelating agent such as EDTA does not necessarily protect a peptide form oxidation, but any possible protection will very much depend on the nature of the involved species (and, of course, the nature of the transition metal). Second, there may be more than one oxidizing species in an EDTA-containing system. In the case of our model system involving Thr-Met and $[FE^{II}(EDTA)]^{2-}$/hydrogen peroxide, these species may be hydroxyl radicals (responsible for acetaldehyde) as well as metal-bound peroxides (responsible for Met sulfoxide). In the following, we have extended these studies to model peptides containing both a Met and a His residue.

Oxidation of His- and Met-containing peptides: Intramolecular catalysis by His

When several model peptides containing His and Met were subjected to oxidation by a system consisting of dithiothreitol, ferric iron, and molecular oxygen, the major oxidation products were the corresponding Met sulfoxide-containing peptides (*37, 41*). Moreover, for some peptides, the presence of His catalyzed the oxidation of Met, indicated by higher oxidation yields of peptides containing both His and Met as compared to peptides that contained only Met (*37, 41*). In the absence of EDTA, the sulfoxide represented ca. 67% of the overall product yield for Gly-His-Gly-Met-Gly-Gly-Gly. In the additional presence of EDTA (or other polyaminocarboxylate chelators), the respective yields of Met sulfoxide decreased. However, this was not due to any particular overall protective effect of EDTA as the overall peptide degradation was still significant in the present of EDTA. Rather, the EDTA-containing system promoted the formation of several other oxidation products besides Met sulfoxide such as 2-oxo-histidine (structure **5**), of which significant yields were formed (*41*). While the formation of 2-oxo-histidine could be inhibited by the addition of 2-propanol, this was not (or significantly less) the case for Met sulfoxide. Thus, also in the dithiothreitol/ferric iron/molecular oxygen-system, the presence of EDTA leads to the formation of different reactive oxygen species promoting a variety of reactions with different oxidation products.

Oxidation of Met-Met: Diastereoselective product formation

We have discussed the importance of equilibria 5 and 6 for the actual fate of intermediates formed by an initiating oxidative event at the sulfur of Met, i.e. electron transfer vs. reaction of hydroxyl radicals. It seemed then appropriate to investigate a peptide in which an intramolecular sulfur-sulfur interaction may lead to a preferred formation of sulfur radical cations, favored through intramolecular stabilization of these cations as [>S∴S<]+ complexes. For Met-Met such intramolecular complexes had been characterized by means of pulse radiolysis (*42*) and laser photolysis (using triplet carboxybenzophenone) (*43*). When the peptide L-Met-L-Met was subjected to one-electron oxidation by triplet carboxybenzophenone in the presence of oxygen, we observed the formation of monosulfoxides and disulfoxides of L-Met-L-Met (*44*). Most interestingly, however, we could characterize the formation of two diastereomeric azasulfonium derivatives (**6a** and **6b**), different at the configuration of the sulfur, and present in an approximate 3:1 ratio (*44*). An absolute assignment of the observed ratio to the two species is currently in progress. These dehydro-Met-Met

products could also be obtained through oxidation of L-Met-L-Met by I_3^-, though with a clean ratio of 1:1. Thus, the one-electron transfer oxidation process displays a measurable diastereoselectivity and it appears that sulfur-sulfur interaction, most probably within an initially formed complex [>S∴S<]+, is an important factor in the observed diastereoselectivity. With this observation, we can provide another interesting example on how peptide sequence (and structure) can influence product formation during peptide oxidation.

Conclusions

The present examples demonstrate the manifold of possible oxidation pathways that can be encountered during the oxidation of peptides even when the nature of the reactive species is known and the systems under study are relatively simple. In a pharmaceutical formulation it will be considerably more difficult to characterize oxidation mechanisms and to design stabilization strategies based on the oxidation mechanisms. Nevertheless, with more experimental work in progress on the oxidation of more complex peptides, we hope to achieve such a goal at some time in the future.

Acknowledgments

The research described herein was supported by the NIH (PO1AG12993-01), the Association For International Cancer Research (AICR), an AFPE fellowship (to F.Z.), a graduate student fellowship from R.W. Johnson, and a Self Fellowship (to B.M.).

Literature cited

(1) Wardman, P. *J. Phys. Chem. Ref. Data* **1989**, *18*, 1637.
(2) Manning, M.C.; Patel, K.; Borchardt, R.T. *Pharm. Res.* **1989**, *6*, 903-918.
(3) Stadtman, E.R. *Annu. Rev. Biochem.* **1993**, *62*, 797-821.
(4) Battersby, J.E.; Mukku, V.R.; Clark, R.G.; Hancock, W.S. *Anal. Chem.* **1995**, *67*, 447-455.
(5) Dean, R.T.; Gieseg, S.; Davies, M.J. *Trends in Biochem. Sci.* **1993**, *18*, 437-441.
(6) Gebicki, S.; Gebicki, J.M. *Biochem. J.* **1993**, *289*, 743-749.
(7) Fu, S.L.; Hick, L.A.; Sheil, M.M.; Dean, R.T. *Free Rad. Biol. Med.* **1995**, *19*, 281-292.
(8) Davies, M.J.; Fu, S.; Dean, R.T. *Biochem. J.* **1995**, *305*, 643-649.
(9) Fu, S.; Gebicki, S.; Jessup, W.; Dean, R.T. *Biochem. J.* **1995**, *311*, 821-827.
(10) Neuzil, J.; Gebicki, J.M.; Stocker, R. *Biochem. J.* **1993**, *293*, 601-606.
(11) Prütz, W.A. In *Sulfur-Centered Reactive Intermediates in Chemistry and Biology*; Chatgilialoglu, C., Asmus, K.-D., Eds.; NATO ASI Series, Plenum Press, New York, **1990**, Vol. 197; pp 389-399.
(12) Stadtman, E.R. *Free Rad. Biol. Med.* **1990**, *9*, 315-325.
(13) Stadtman, E.R.; Oliver, C.N. *J. Biol. Chem.* **1991**, *266*, 2005-2008.
(14) Davies, M.J.; Gilbert, B.C.; Haywood, R.M. *Free Rad. Res. Comms.* **1991**, *15*, 111-127.
(15) Encinas, M.V.; Lissi, E.A.; Vasquez, M.; Olea, A.F.; Silva, E. *Photochem. Photobiol.* **1989**, *49*, 557-563.
(16) Timmins, G.S.; Davies, M.J. *J. Photochem. Photobiol.: B. Biol.* **1993**, *21*, 167-173.
(17) Timmins, G.S.; Davies, M.J. *J. Photochem. Photobiol.: B. Biol.* **1994**, *24*, 117-122.
(18) Garrison, W.M. *Chem. Rev.* **1987**, *87*, 381-398.
(19) Franzini, E.; Sellak, H.; Hakim, J.; Pasquier, C. *Biochim. Biophys. Acta* **1993**, *1203*, 11-17.
(20) Felix, K.; Lengfelder, E.; Hartmann, H.-J.; Weser, U. *Biochim. Biophys. Acta* **1993**, *1203*, 104-108.
(21) Walsh, M.P.; Stevens, F.C. *Biochemistry* **1977**, *16*, 2742-2749.
(22) Guerini, D.; Krebs, J.; Carafoli, E. *Eur. J. Biochem.* **1987**, *170*, 35-42.
(23) Yao, Y.; Yin, D.; Jas, G.; Kuczera, K.; Williams, T.D.; Schöneich, Ch.; Squier, T.C. *Biochemistry* **1996**, *35*, 2767-2787.
(24) Hühmer, A.F.R.; Gerber, N.C.; Ortiz de Montellano, P.R.; Schöneich, Ch. *Chem. Res. Toxicol.* **1996**, *9*, 484-491.

(25) Michaelis, M.L.; Bigelow, D.J.; Schöneich, Ch.; Williams, T.D.; Ramonda, L.; Yin, D.; Hühmer, A.F.R.; Yao, Y.; Squier, T.C. *Life Sci.* **1996**, *59*, 405-412.
(26) Schöneich, Ch.; Zhao, F.; Madden, K.P.; Bobrowski, K. *J. Am. Chem. Soc.* **1994**, *116*, 4641-4652.
(27) Miller, B.L.; Schöneich, Ch., unpublished results.
(28) Bobrowski, K.; Schöneich, Ch.; Holcman, J.; Asmus, K.-D. *J. Chem. Soc. Perkin Trans. 2* **1991**, 975-980.
(29) Bobrowski, K.; Schöneich, Ch. *Radiat. Phys. Chem.* **1996**, *47*, 507-510.
(30) Bobrowski, K.; Schöneich, Ch. *J. Chem. Soc. Chem. Comm.* **1993**, 795-797.
(31) Schöneich, Ch.; Bobrowski, K. *J. Phys. Chem.* **1994**, *98*, 12613-12620.
(32) Razskazovskii, Y.; Sevilla, M.D. *J. Phys. Chem.* **1996**, *100*, 4090-4096.
(33) Bonifacic, M.; Weiss, J.; Chaudhri, S.A,; Asmus, K.-D. *J. Phys. Chem.* **1985**, *89*, 3910-3914.
(34) Clark, T. In *Sulfur-Centered Reactive Intermediates in Chemistry and Biology*; Chatgilialoglu, C., Asmus, K.-D., Eds.; NATO ASI Series, Plenum Press, New York, **1990**, Vol. 197; pp 13-18.
(35) Schöneich, Ch.; Yang, J. *J. Chem. Soc. Perkin Trans. 2* **1996**, 915-924.
(36) Bobrowski, K.; Schöneich, Ch.; Holcman, J.; Asmus, K.-D. *J. Chem. Soc. Perkin Trans. 2* **1991**, 353–362.
(37) Schöneich, Ch.; Zhao, F.; Wilson, G.S.; Borchardt, R.T. *Biochim. Biophys. Acta* **1993**, *1158*, 307-322.
(38) Li, S.; Schöneich, Ch.; Wilson, G.S.; Borchardt, R.T. *Pharm. Res.* **1993**, *10*, 1572-1579.
(39) Zhao, F.; Ghezzo-Schöneich, E.; Aced, G.; Milby, T.; Schöneich, Ch., submitted to *Biochemistry*.
(40) Schöneich, Ch.; Aced, A.; Asmus, K.-D. *J. Am. Chem. Soc.* **1993**, *115*, 11376-11383.
(41) Zhao, F.; Yang, J.; Schöneich, Ch. *Pharm. Res.* **1996**, *13*, 931-938.
(42) Bobrowski, K.; Holcman, J. *J. Phys. Chem.* **1989**, *93*, 6381-6387.
(43) Hug, G.L.; Marciniak, B.; Bobrowski, K. *J. Photochem. Photobiol. A* **1996**, *95*, 81-88
(44) Miller, B.L.; Schöneich, Ch., manuscript in preparation.

Chapter 5

A Discussion of Limitations on the Use of Polymers for Stabilization of Proteins During the Freezing Portion of Lyophilization

David M. Barbieri[1], Martin C. Heller[1], Theodore W. Randolph[1,3], and John F. Carpenter[2]

[1]Department of Chemical Engineering, Campus Box 424, ECCH 111, University of Colorado, Boulder, CO 80309-0424
[2]Department of Pharmaceutical Sciences, School of Pharmacy, University of Colorado Health Sciences Center, Denver, CO 80262

A thermodynamic model based on the theoretical framework of Timasheff and coworkers has been developed to consider the protein stabilization offered by polymeric co-solvents. Inspection of such systems reveals that the large transfer free energies (and presumably protein stability) rendered by polymeric excipients such as poly(ethylene glycol) increase with increasing polymer concentration. These same polymers, however, commonly induce phase splits in aqueous solutions, presenting limitations to the protection conferred. Further consideration of freeze-drying formulations suggest that such phase splits are a likely consequence of the concentrating effects of freezing aqueous solutions. Experimental studies of hemoglobin lyophilized in poly(ethylene glycol) / dextran mixtures give evidence that liquid / liquid phase separation *per se* occurring during the course of the lyophilization cycle can have detrimental effects on the structural integrity of protein in the dried state.

Protein-based pharmaceuticals are becoming more prevalent with advances in recombinant technologies. Concomitantly, formulations for these pharmaceuticals must be developed to insure that a native, functional protein is delivered to the patient. Lyophilization is a popular last step in the production of recombinant pharmaceuticals as a properly lyophilized protein is often more stable than liquid formulations. The lyophilization process is, however, inherently harsh on these proteins which are only marginally stable (*1*), even in their cellular environment.

There are additives that are known to help stabilize pharmaceutical proteins against denaturation and aggregation, many already possessing FDA approval for use in parenteral formulations. Among these additives are polymeric materials such as

[3]Corresponding author

poly(ethylene glycol) (PEG) and dextran. A number of these polymers, PEG in particular, have been shown to be potent cryoprotectants, but they are also desirable freeze-drying components because of their typically high glass transitions and bulking properties (2). Coincidentally, the same polymeric materials are also common constituents in aqueous two-phase extraction systems used for, among other things, protein separation and purification.

An interesting question is raised. Can phase separation occur during lyophilization and, if so, are there any consequences regarding protein stability? If phase separation does occur, then a protein would be free to partition into either phase. At the very least, the composition of that phase may be very different than the original formulation. If the protein partitions into a phase deficient in stabilizing excipients, then it would be expected that any stabilizing effects would be diminished.

In this chapter, we consider a system of hen egg-white (HEW) lysozyme with PEG 3350 and dextran T70 in the context of the theoretical framework developed by Timasheff and co-workers (3-15) to describe the stabilization of labile proteins in aqueous solutions. The same theory has been effectively extrapolated to describe the stabilization of proteins during freezing (16). Structural data for recombinant hemoglobin lyophilized in a similar PEG-dextran system has been examined, giving experimental evidence of the structural damage introduced by a phase split (17).

Phase Splitting and the Lyophilization Process

Liquid / liquid phase separation results when the repulsive interactions between molecules overcome the entropic energy gain of mixing, and the energetically most favorable state of the system becomes one where unlike molecules are separated. This results in a physical "split" of the system into two phases, with the large scale separation being driven by density differences in the two phases. Phase splitting during lyophilization can have dramatic effects on the state of the components involved. The systems considered here are two-polymer systems of PEG and dextran. While this is an unlikely combination for a protein formulation, it is an attractive model to consider theoretically and experimentally, and can lead to valuable insights into the effects of a phase split on a protein formulation. Indeed, PEGs are known to form two-phase aqueous systems with salts as well as with other polymers that are more common constituents in protein formulations, making phase separation a true industrial concern.

Inspection of the lyophilization process reveals that phase splitting is a highly likely possibility. While protein formulations typically do not start at concentrations where phase separation will occur, the concentrating effects of freezing can force them into a two phase region. As the temperature of a polymer/protein aqueous electrolyte solution is lowered, the first major change is the formation of ice at some subzero temperature determined by the colligative properties of the solution. This often occurs after the sample has been supercooled to temperatures below the equilibrium freezing point. A definite result of the bulk formation of pure ice is the concentration of components in the remaining liquid phase. At some point during

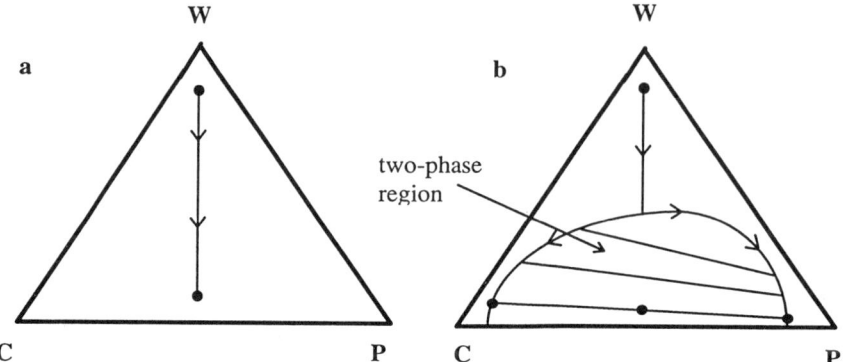

Figure 1. Tertiary phase diagram of a representative protein formulation showing the concentrating effect of the freezing of pure ice. Components are: W - water, C - co-solvent, P - polymer. Diagram b) shows the freeze concentration phenomenon with the occurrence of a phase split.

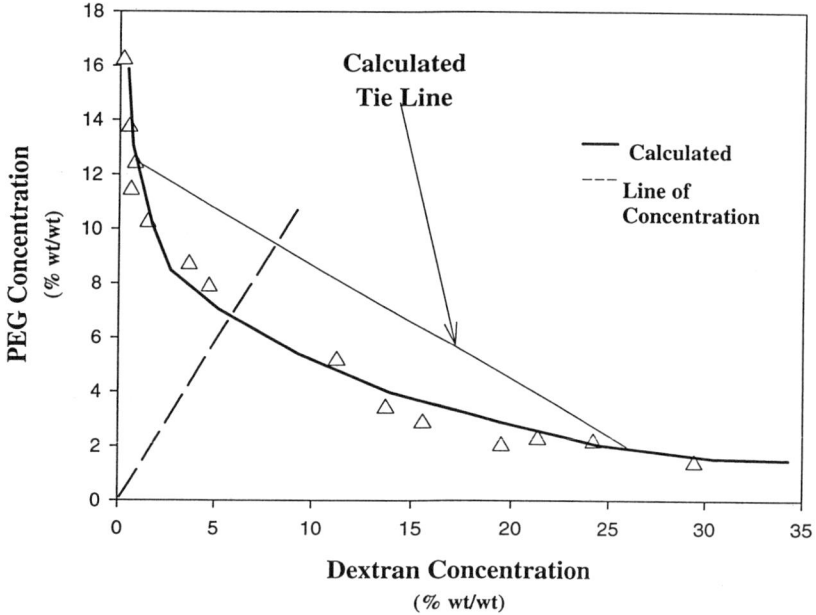

Figure 2. Phase diagram showing the two-phase region for a PEG 3350 / dextran T70 system. Solid lines calculated using model of King *et al.* (see Appendix A). Triangles are experimental data points from Albertsson (*22*). The dashed line represents the path of concentration that the non-ice components of an example aqueous system would follow during the freezing of pure ice. The two phase region lies above the binodal curve.

this freeze-concentration, a liquid / liquid phase split may occur. This phenomenon of freeze concentration with and without a phase split is shown in Figure 1.

Without a phase split, freeze-concentration would continue until the system is cooled below the glass transition temperature of the cosolvent. At this point, a glassy phase would form and essentially fix the concentration of the non-ice phase. If a phase split occurs, the situation is more complex.

As mentioned above, the protein may partition into either phase. The shape of the two-phase region and its tie-lines as determined by thermodynamics would dictate how much of each phase exists and what the actual phase compositions are. To complicate the situation, there may now be two different glass transition temperatures (*18*).

Figure 2 shows a phase diagram for PEG 3350 and dextran T70 calculated using the model and parameters presented by King *et al.* (*19*) (see Appendix). As depicted by the dashed line, the path on which the non-ice fraction of the system travels during freeze concentration can easily cross into a two-phase region. The calculated tie line indicates the composition of the two phases that are formed, one which is enriched in dextran, and the other in PEG. Since such freeze concentration will arise during the freezing step of lyophilization, it seems that this process would be conducive to liquid / liquid phase-splitting, even if the initial formulation is far from the two-phase region.

Mechanism of Protection

The mechanism used to describe the phenomenon of protein stabilization in solution by excipients is detailed elsewhere in the literature (*3-15*). The following is a brief overview to set the stage for consideration of phase separation during lyophilization. Consider a multicomponent system. Component 1 is given as the solvent. Water is typical, although a pseudo solvent such as an aqueous buffer solution may also be considered component 1. The latter is actually more common as many proteins are solvated in the presence of buffer salts. The protein is denoted as component 2. Component 3 is taken to be the excipient, or cosolvent, of the system.

There are three thermodynamic quantities that are used to describe the stabilization of proteins by excipients. The one of interest for the purposes of this discussion is called the transfer free energy. It is the change in free energy of the protein (μ_2) on transfer from pure water (pseudo solvent) to a cosolvent system. Mathematically it can be expressed as

$$\Delta \mu_{2,tr}^{N} = \mu_{2,c}^{N} - \mu_{2,w}^{N} \tag{1}$$

The superscript N denotes the native conformation of the protein (a D superscript denotes the denatured, or unfolded, state). Subscripts 2, c, and w denote component 2 (protein), cosolvent system, and pure water system, respectively.

Figure 3 depicts a typical free energy diagram for a protein and a cryoprotectant cosolvent. As shown in Figure 3, the transfer free energy for a cryoprotectant will, in general, be positive. The quantity, $\Delta G_{denaturation}$, determines the relative amounts of protein in the native and denatured forms at equilibrium. A larger $\Delta G_{denaturation}$ would correspond to a more stable protein -- relatively more protein in the native state than the denatured state. Since, for a cryoprotectant,

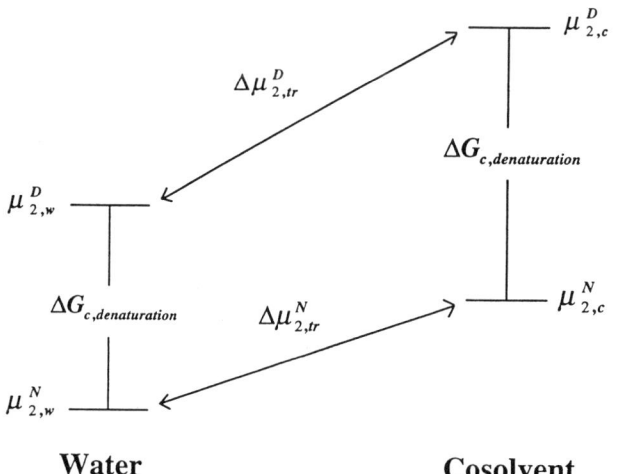

Figure 3. Energy schematic depicting the effects of a typical protein stabilizer. The transfer free energy for a protein into a stabilizing cosolvent will be positive for both the native and denatured forms. The denatured form will have a larger transfer free energy than the native form, resulting in a larger $\Delta G_{denaturation}$ in the cosolvent system. Thus, more protein will be in the native state than the denatured state in the presence of the cosolvent, concurring a relative stability. Symbols are described in the text. Adapted from (14). At low temperatures, PEG is excluded from protein surfaces, acting as a stabilizer as shown here (23,24). It should be noted, however, that at high temperatures, PEG can bind to proteins, acting as a denaturant (12,25), in which case this figure is no longer valid.

$\Delta\mu^D_{2,tr} > \Delta\mu^N_{2,tr}$, $\Delta G_{denaturation}$ will be larger in the cosolvent system than in the pure water system, i.e. a relative stabilization has been conferred to the protein.

It has been demonstrated that excipients which are preferentially excluded from the surface of the protein, such as polymeric materials, will confer stability to proteins through the above mechanism (*3-15*). Globular proteins are assumed to form a compact native structure in solution. An increase in surface area upon unfolding is given as a physical rationale for the larger transfer free energy for the denatured state than for the native state (*12*). From this argument it would seem that the transfer free energy is a natural yardstick to measure the cryoprotectant effectiveness of an excipient. A larger transfer free energy should then correspond to a better protectant.

Chemical Potential Relief Valve?

From the mechanism outlined above, it seems than a large transfer free energy is desirable in stabilizing a protein. If adding more and more of a polymeric excipient will increase the transfer free energy, then it is probably a good idea to add as much polymer as practical. Figure 4 shows the transfer free energy for lysozyme as a function of PEG concentration. Transfer free energies as a function of polymer concentration are calculated using a model presented by King *et al.* (*19*) (see Appendix). Indeed, the transfer free energy does increase with increasing polymer concentration. Adding increasing quantities of polymer, however, will make phase separation even more likely.

Will a phase separation act as a chemical potential relief valve? Phase splits occur because the separation of unlike molecules leads to a lower free energy system than the single phase system. Thus, the possibility of a phase split raises a question of diminishing returns; in essence the phase split acts as a chemical potential "relief valve" in the protein stabilizing system. There is a possibility that some protection -- increased transfer free energy -- may even be lost. At the very least, subsequent addition of excipient will not provide the same increase in protection.

Figure 5 shows the transfer free energy of lysozyme in a PEG-dextran system with and without a phase split, calculated using the model presented by King *et al.* (*19*). Indeed, there is an attenuation of transfer free energy, and presumably protein protection, realized after a phase split occurs. The occurrence of such an attenuation of protection has large implications in the design of a freeze drying formulation. There is clearly a limit in the amount of "protection" as represented by the preferential exclusion principle that can be gained by the addition of a polymeric excipient when a phase split can occur. Also, if the phase split results in a protein-rich phase with minimal excipient (protein partitions away from intended stabilizer), then the excipient would not be availiable to protect the protein during subsequent drying. Finally, the advent of a phase separation *per se* can have detrimental effects on protein structural stability as we will see in the following sections.

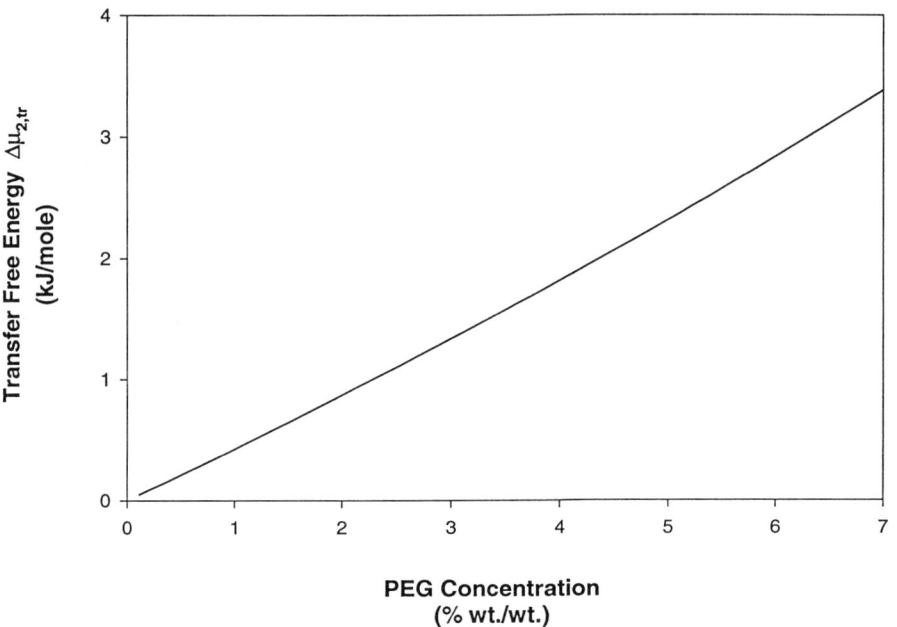

Figure 4. Calculated transfer free energy of hen egg-white lysozyme as a function of PEG 3350 concentration. Transfer free energy can be considered a measure of the protectant effectiveness of an excipient. Details of calculations can be found in Appendix B.

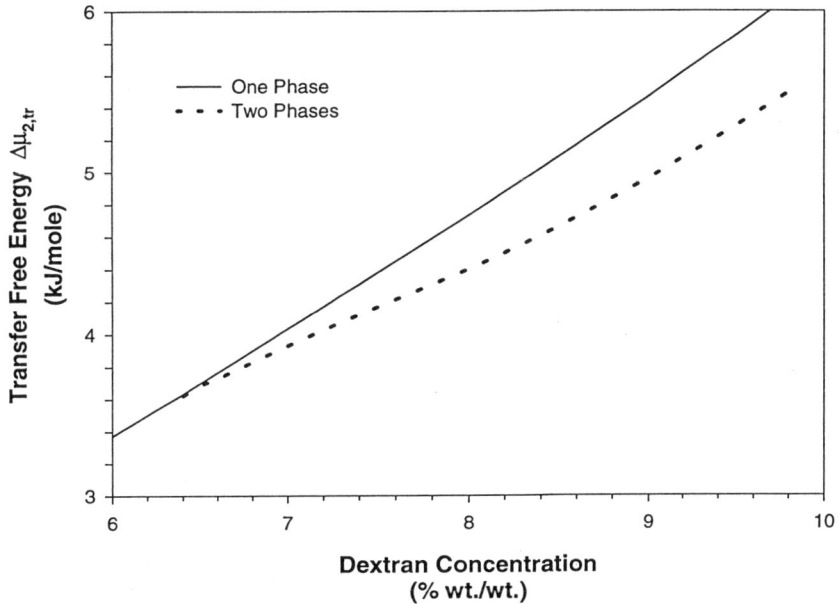

Figure 5. Calculated transfer free energy of lysozyme in a PEG 3350 / dextran T70 system with (dashed line) and without (solid line) the onset of a phase split. PEG concentration is maintained constant at 7%. Details of calculations can be found in Appendix B.

Structural Information

Fourier Transform Infrared Spectroscopy. Fourier transform infrared (FTIR) spectroscopy has proven an essential tool in the study of protein stabilization primarily because it allows the study to be performed in any state (i.e., aqueous, frozen, dried, or even insoluble aggregate). Dong *et al.* (20) have recently published a review on the use of FTIR for the study of protein stability during lyophilization which offers an excellent introduction to the technique. While the vibrational spectra of protein molecules exhibit a number of characteristic frequency bands, the amide I band (1700-1600 cm^{-1}) has seen the widest use in assessing protein structure. This band is due primarily to the C=O stretching vibrations of the peptide linkages that make up the backbone structure of proteins, and is thus sensitive to small variations in molecular geometry. Each type of secondary structure (α-helix, β-sheet, etc.), in principle, gives rise to a different C=O stretching frequency in the amide I region of the IR spectrum, although, because of the broad bandwidth of these components, the amide I absorbance usually appears as a single composite signal. Mathematical band-narrowing techniques have been developed to allow the resolution of underlying amide I peaks and thus qualitative and often quantitative information about the protein's secondary structure. Of the techniques developed, we have found the second derivative method of analysis most useful primarily because it is not dependent on the subjective input of arbitrary parameters (see (20) for a thorough description and comparison of band narrowing techniques).

Experimental Methods. Lyophilization experiments were performed on samples of recombinant hemoglobin (Somatogen, Inc., Boulder, CO) to determine the effects of polymers on the retention of native structure in the dried state. All samples were buffered in 5 mM phosphate, pH 7.4, and contained 150 mM NaCl. Sample volumes of 1 mL were aliquoted in 5 mL vials. Freeze drying was performed using a FTS Systems microprocessor controlled tray dryer with the following conservative cycle: freezing with shelf temperature set at -50°C for 120 minutes; primary drying with shelf temperature at -20°C and vacuum set at 60 mT for 1000 minutes; secondary drying steps of 0°C, 60 mT for 120 minutes, 10°C, 60 mT for 120 minutes, and 25°C, 60 mT for 120 minutes. Dried cakes were solid plugs with no apparent meltback. FTIR spectra of aqueous and dried samples were collected using a Nicolet Magna Model 550 spectrometer equipped with a dTGS detector. Aqueous solutions were placed into a liquid cell (Beckman FH-01) in which CaF_2 windows are separated by a 6 μm spacer. For dried samples, KBr pellets of approximately 1-5 mg of lyophilized protein mixture and 200 mg KBr were pressed and placed immediately in the IR sample chamber in a magnetic sample holder. Spectra of amorphous two phase systems were taken on ground and well mixed cakes, and thus are representative of a combination of material in the two phases. A 256-scan interferogram was acquired in single beam mode with a 4 cm^{-1} resolution. Buffer and water vapor components were subtracted using Nicolet software. A seven-point smoothing was performed on the second derivative spectra to remove possible white noise, and a baseline correction was performed over the amide I region. All spectra were normalized by their total area between 1700 cm^{-1} and 1600 cm^{-1} to allow direct comparison.

Results: Lyophilization Studies. Figure 6 shows the amide I region of the second derivative FTIR spectra of hemoglobin both in solution and lyophilized in the presence of both PEG 3350 and dextran T500. Hemoglobin is primarily an α-helix protein (87% by x-ray crystallography (21)); this is reflected in the dominant α-helix IR band near 1656 cm^{-1} seen in the spectrum of hemoglobin in solution, which is considered the native secondary structure. Clearly, there is a loss of intensity of this α-helix band in the spectrum of hemoglobin lyophilized in buffer alone, indicating a loss of native structure, or a destabilization, that results from freeze drying. Spectra of samples lyophilized with increasing levels of PEG, however, show a progressive return in native structure (increase in α-helix content). This result is congruent with the native state transfer free energy calculations shown in Figure 4. That is, increasing polymer concentration confers added stability to the protein. This general trend is not true with all polymers, however, at least not to the same extent. While the spectra of samples containing dextran (figure 6b) show some stabilizing effects on hemoglobin's structure, the effect is not as dramatic as that seen with PEG.

As was mentioned earlier, aqueous PEG/ dextran mixtures will undergo a phase split at appropriate concentrations. Systems were chosen from known phase diagrams (22) to ascertain experimentally the effects that a phase split occurring during lyophilization has on hemoglobin structural stability. Figure 7 shows the α-helix band depth of the second derivative FTIR spectra of hemoglobin samples lyophilized in various PEG /dextran mixtures. Since the α-helix band is considered the primary indicator of native structure in these hemoglobin experiments, results have been focused on this band. Thus, the liquid control is considered the native "α-helix band depth" or α-helix content. Mixtures of 7% PEG / 7% dextran undergo a definite phase split at room temperature. The PEG rich (top) phase is approximately 10% PEG / 0.2% dextran, while the dextran rich (bottom) phase is 20% dextran / 2% PEG. Hemoglobin partitions in this system such that the hemoglobin concentration in the dextran rich phase is about three times that in the PEG rich phase. Protein at the concentrations used, however, has little effect on the PEG / dextran phase diagram (22). Lyophilized samples at this concentration are clearly two phase before freezing and drying. As can be seen by the α-helix band depth in Figure 7, the system offers some protection over lyophilizing in buffer alone. The 5% PEG / 5% dextran system sits nearly on the two phase boundary at 0°C and thus also undergoes a phase separation prior to freezing. The α-helix content in the dried state is nearly identical to the 7% system. Concentrations of 4% PEG / 4% dextran, however, are well below the two phase boundary even at 0°C. Phase separation occurring in this sample (as is evident in the scanning electron microscopy of the dried cake, data not shown) happens as a result of freeze concentration during the lyophilization cycle. The IR spectrum of the dried state hemoglobin in the 4% PEG / 4% dextran system reveals a clear decrease in α-helix content over hemoglobin freeze dried in buffer alone. This result indicates that there is a detrimental effect on the protein as a result of the phase split occurring during the course of the lyophilization cycle. When a phase separation occurs, one phase nucleates as dispersed droplets in a continuum of the other phase. At the sub-zero temperatures where this would occur during freezing of a lyophilization sample, high viscosity will hinder the kinetics of the macroscopic phase separation, that is the migration and coalescence of the dispersed

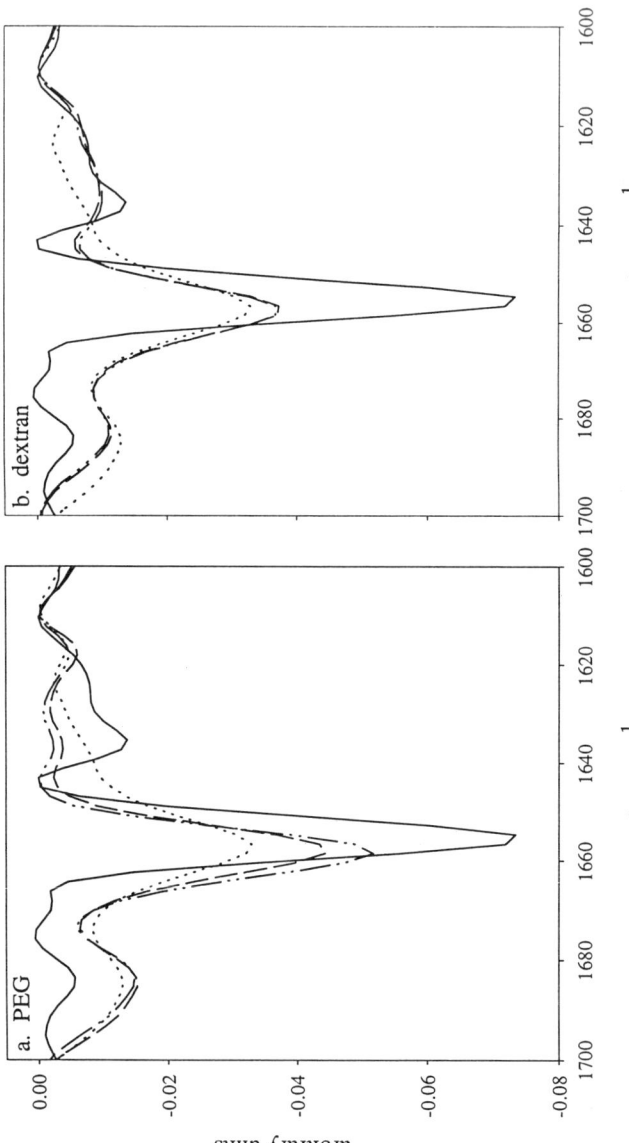

Figure 6. Amide I region of the second derivative FTIR spectra of hemoglobin in solution and lyophilized in various concentrations of: a) PEG 3350, and b) dextran T500. Solid curve - hemoglobin in liquid solution, Dotted curve - dried state hemoglobin lyophilized in buffer alone, dashed curve - dried state hemoglobin lyophilized with 5% polymer, dash-dot-dot curve - dried state hemoglobin lyophilized with 10% polymer.

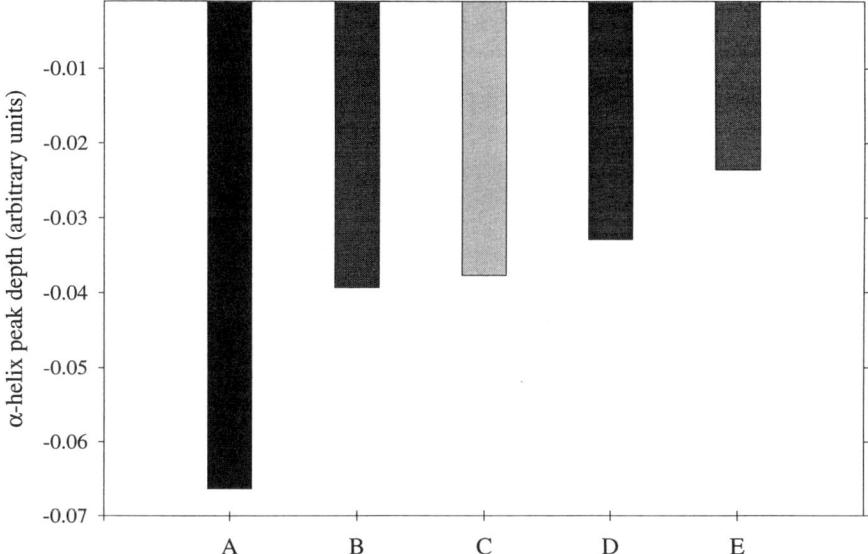

Figure 7. α-helix peak depth (~1656 cm^{-1}) from second derivative FTIR spectra of dried state hemoglobin. A - hemoglobin in solution (native state reference). B - lyophilized with 7% PEG 3350 / 7% dextran T500. C - lyophilized with 5% PEG / 5% dextran. D - lyophilized in buffer only. E - lyophilized in 4% PEG / 4% dextran. Modified from (17).

drops. When the sample reaches a glass transition, the dispersed phase is locked from further separation, leaving an enormous excess of interfacial area. We speculate that this increase in interfacial surface and the added stresses it presents are responsible for the observed protein damage.

Conclusions

While polymeric excipients are desirable stabilizers in protein formulations due to their typically large transfer free energies, there are definite limitations. Excipients such as poly(ethylene glycol) confer increasing transfer free energies with increases in concentration. These same polymers, however, are often incompatible with other formulation components at high concentrations and can cause a phase split to occur. At the very least, the onset of a phase split reduces the effectiveness of the stabilizer by providing a chemical potential "relief valve." Further, because of the large concentrating effects of freezing pure ice out of solution during lyophilization, phase splits can occur at relatively low initial polymer concentrations. It has been demonstrated experimentally that phase separation *per se* occurring during the lyophilization cycle can actually have detrimental effects on the structural stability of a protein.

Appendix A.

King, Blanch, and Prausnitz in (19) present a thermodynamic model for the prediction of aqueous phase separations. It is based on a osmotic virial expansion truncated after the second-virial-coefficient term. Chemical potentials of polymer (2) and polymer (3) as a function of molality are given by:

$$\Delta\mu_2 = RT(\ln m_2 + a_{22}m_2 + a_{23}m_3) \quad (2)$$

$$\Delta\mu_3 = RT(\ln m_3 + a_{33}m_3 + a_{23}m_2) \quad (3)$$

where m_i is the molal concentration of polymer i, and a_{ii} is the interaction coefficient, which can be related to traditional second virial coefficients, A_{ii}, obtainable by low-angle light scattering, by:

$$2A_{22} = 1,000\frac{a_{22}}{(M_2)^2} \quad (4a)$$

$$2A_{33} = 1,000\frac{a_{33}}{(M_3)^2} \quad (4b)$$

$$2A_{23} = 1,000\frac{a_{23}}{M_2 M_3} \quad (4c)$$

Here, M_i is the molecular weight. The chemical potential for water (1) can then be calculated from the Gibbs-Duhem relation:

$$\Delta\mu_1 = -RTV_1\rho_1\left(m_2 + m_3 + \frac{a_{22}}{2}(m_2)^2 + \frac{a_{33}}{2}(m_3)^2 + a_{23}m_2m_3\right) \quad (5)$$

where V_1 and ρ_1 are the molal volume and density of water (or pseudo-solvent), respectively. A phase diagram can be determined with these equations, the appropriate mass balance equations, and the following conditions for equilibrium:

$$\mu_1' = \mu_1'' \quad (6a)$$

$$\mu_2' = \mu_2'' \quad (6a)$$

$$\mu_3' = \mu_3'' \quad (6a)$$

Appendix B.

For calculation of protein chemical potentials and transfer free energies, the more detailed model of Haynes, et al. (26) was used. This mathematical model describes the protein's chemical potential as a function of cosolvent concentration, enabling the transfer free energy to be obtained:

$$\Delta\mu_{2,tr} \equiv \int_0^{m_3}\left(\frac{\partial\mu_2}{\partial m_3}\right)_{T,P,m_2} dm_3 \quad (7)$$

where m_i is the molality of compound i, T is the temperature, and P is the pressure.

Details of the model (as described for aqueous two-phase partitioning calculations) are given elsewhere (26), but an overview is given here:

The modified Helmholtz energy is first divided into ideal and excess parts as shown in equation (4)

$$A' = A'_{id} + A'_{ex} \tag{8}$$

The chemical potential of any species in the system is simply given by its partial derivative of A':

$$\mu_j = \left(\frac{\partial A'}{\partial n_j}\right)_{\mu_o, n_i(i \neq j, 0), T, V} \tag{9}$$

where n_j is the number of moles of species j. The chemical potential can similarly be divided up into its ideal and excess parts which correspond to the ideal and excess parts of A'.

$$\mu_j = \mu_{j,id} + \mu_{j,ex} = \left(\frac{\partial A'_{id}}{\partial n_j}\right)_{\mu_o, n_i(i \neq j, 0), T, V} + \left(\frac{\partial A'_{ex}}{\partial n_j}\right)_{\mu_o, n_i(i \neq j, 0), T, V} \tag{10}$$

After transformation from the modified Helmholtz framework (independent variables $m_0, n_{j(j \neq 0)}, T, V$) to the Gibbs (experimental) framework (independent variables n_j, T, P), the final expression for the chemical potential of species i is given as (27)

$$\mu_i = \mu_i^\theta + RT \ln \phi_i + RT\left(1 - \frac{\overline{V}_i^\theta \sum_{j=0}^{m} n_j}{\sum_{k=0}^{m} \overline{V}_k^\theta n_k}\right) - \overline{V}_i^\theta P_{EX} + \mu_{i,EX} \tag{11}$$

Here μ_i^θ is the standard state chemical potential of species i, R is the universal gas constant, T is the absolute temperature, ϕ_i is the volume fraction of species i, $\overline{V}_i^\theta$ is the partial molar volume of species i, n_j is the number of moles of species j, and P_{EX} is the excess pressure.

The first three terms on the right-hand side of equation (11) comprise the well-defined ideal part of the chemical potential. The excess pressure term arises on transformation from the A' framework to the Gibbs framework, and is given as

$$P_{ex} = -\left(\frac{\partial A'_{ex}}{\partial V}\right)_{\mu_0, n_{j \neq 0}, T} \tag{12}.$$

The last term is the excess chemical potential as defined in equation (10):

$$\mu_{j,ex} = \left(\frac{\partial A'_{ex}}{\partial n_j}\right)_{\mu_o, n_i(i \neq j, 0), T, V} \tag{13}.$$

The excess Helmholtz free energy A'_{ex} is considered to be composed of five contributions:

$$A'_{ex} = A'_{ex,hs} + A'_{ex,na} + A'_{ex,ic} + A'_{ex,cc} + A'_{ex,ve} \tag{14}$$

where $A'_{ex,hs}$ (hard-sphere) is the contribution to the excess modified Helmholtz free energy from combining hard spheres of unequal sizes in a continuum:

$$A'_{Ex,hs} = A'_{hs} - A'_{id} \tag{14a}$$

$$= -RT\sum_i n_i \ln(1-\xi_3) - \frac{RT}{\overline{V}_o^\theta} \sum_{i \neq 0} n_i \overline{V}_o^\theta$$

$$- \frac{RTV}{\overline{V}_o^\theta}\left(1-\sum_{i \neq 0}\phi_i\right)\ln\left(1-\sum_{i \neq 0}\phi_i\right)$$

$$+ \frac{6kTV}{\pi}\left[\frac{3\xi_1\xi_2 - (\xi_2)^3/\xi_3}{1-\xi_3} + \frac{(\xi_2)^3}{(\xi_3)^2}\ln(1-\xi_3) + \frac{(\xi_2)^3}{(1-\xi_3)^2}\right],$$

where $\overline{V}_o^\theta$ is the solvent partial molar volume, $\xi'_m = V\xi_m = \dfrac{\pi N_{Av}\sum_i n_i d_i^m}{6}$ with $m=0, 1, 2,$ or 3, and d_i is the hard-sphere diameter of species i.

$A'_{ex,na}$ (non-additive) is a first-order correction to the hard-sphere term accounting for penetration of protein macroions into polymers:

$$A'_{Ex,na} = A'_{na} - A'_{hs} = \frac{2\pi RT}{V}\sum_{i \neq 0}\sum_{j \neq 0} n_i \frac{n_j}{\sum_{k \neq 0} n_k} d_{ij}^2 g_{ij}(d_{ij})(\sigma_{ij} - d_{ij}) \tag{14b}$$

where σ_{ij} is the actual closest distance of approach of components i and j, and the pair correlation function g_{ij} for additive hard spheres is taken from Grundke and Henderson (28):

$$g_{ij}(d_{ij}) = \frac{1}{1-\xi_3} + \frac{3\xi_2}{2(1-\xi_3)^2}\left(\frac{d_i d_j}{d_{ij}}\right) + \frac{\xi_2^2}{2(1-\xi_3)^3}\left(\frac{d_i d_j}{d_{ij}}\right)^2 \tag{14c}$$

$A'_{ex,ic}$ (ion charging) is a contribution from charging neutral spheres in a dielectric continuum,

$$A'_{Ex,ic} = \frac{N_{Av}e^2}{\varepsilon}\sum_i \frac{n_i z_i^2}{d_i} \tag{14d}$$

where N_{Av} is Avogadro's number, ε is the permittivity, and z_i is the charge on species i.

$A'_{ex,cc}$ (charge-charge) accounts for charge-charge interactions:

$$\frac{A'_{Ex,cc}}{V} = -\frac{e^2}{\varepsilon}\left(\Gamma\sum_i\frac{\rho_i z_i^2}{1+d_i\Gamma} + \frac{\pi}{2\Delta}\Omega P_n^2\right) + \frac{\Gamma^3 kT}{3\pi}; \tag{14e}$$

$$4\Gamma^2 = \frac{\kappa^2\sum_i \rho_i(1+d_i\Gamma)^{-2}(z_i - \pi d_i^2 P_n/2\Delta)^2}{\sum_i \rho_i z_i^2}$$

$$P_n = \frac{1}{\Omega}\sum_j \frac{d_j\rho_j z_j}{1+d_j\Gamma} \text{ and } \Omega = 1 + \frac{\pi}{2\Delta}\sum_j \frac{\rho_j d_j^3}{1+d_j\Gamma}$$

$$\Delta = 1 - \frac{\pi}{6}\sum_j \rho_j d_j^3 \ , \ \kappa^2 = \frac{4\pi e^2}{\varepsilon kT}\sum_j \rho_j z_j^2 \ , \text{and } \rho_j \text{ is the number density of solute}$$

i.

Finally, $A'_{ex,ve}$ (virial expansion) takes short-range interactions between solutes into account and is calculated from an osmotic virial expansion truncated at the second term:

$$A'_{ex,ve} = RTN_{Av}\sum_{i\neq 0}\sum_{j\neq 0}n_i\frac{n_j}{V}\beta_{ij}^*(\mu_o,T) \tag{14f}$$

β^*_{ij} is the residual second osmotic virial coefficient, which can be calculated from the second osmotic virial coefficient B_{ij} by:

$$\beta_{ij}^* = B_{ij}\frac{MW_i MW_j}{1000 N_{Av}} - 1000\frac{2\pi\sigma_{ij}^3}{3} \tag{15a}$$

or from the solute-solute specific interaction coefficient βij by

$$\beta_{ij}^* = \frac{\beta_{ij}}{2N_{Av}} - 1000\frac{2\pi\sigma_{ij}^3}{3} \tag{15b}$$

More detailed explanations of each of the expressions for each of the aforementioned contributions to A'_{ex} are given by Haynes *et al.*, *(26)*.

Application of the above framework to a specific system requires several parameters. They include: hard-sphere diameters of each solute, solvent molar volume, solute partial molar volumes at infinite dilution, ion charges, solvent dielectric constant, and osmotic second virial coefficient data. With the exception of the osmotic virial coefficient (required for the $A'_{ex,ve}$ term), these parameters are available in the literature or would be routinely measured for a new protein system; required light scattering experiments for the osmotic virial coefficients are relatively straightforward.

Salt parameters are readily available in the literature *(29)*. For application to a particular system, polymer and protein parameters would have to be determined. The charges of proteins can be determined from acid-base titration *(30-35)*. Hard-sphere diameters of polymers can be regressed from differential solvent vapor pressure data *(36)* and protein diameters can be determined from the method of Tyn and Gusek *(37)*. Distances of closest approach are taken as the arithmetic average of species diameters. Protein and polymer partial molar volumes can be regressed from density measurements *(38)* or estimated from correlations *(39)*. Osmotic second virial coefficients can be regressed from low-angle laser-light scattering (LALLS) measurements combined with Raleigh theory *(40,41)*. For the examples presented here, we used the data found in Tables I-IV, which were obtained from data compiled by Haynes, et al., *(26)*.

Table I. Hard sphere diameters d_{ij}

	Diameter, Å
Lysozyme	30.4
Dextran T70	46.8
PEG3350	24.6
K^+	3.04
$H_2PO_4^-$	4.44
HPO_4^{2-}	3.82

SOURCE: Adapted from ref. 26

Table II. Distance of closest approach, σ_{ij}, Å.

	Lysozyme	DextranT70	PEG 3350	K^+	$H_2PO_4^-$	HPO_4^{2-}
Lysozyme	30.4					
Dextran T-70	24.4	33.6				
PEG3350	24.6	------	22.9			
K^+	16.7	17.6	13.2	3.04		
$H_2PO_4^-$	17.4	18.4	13.8	3.74	4.44	
HPO_4^{2-}	17.1	18.6	13.8	3.43	4.13	3.82

For charged species, there is assumed to be no intermolecular penetration, and σ_{ij} may be obtained from Table as $(d_{ii}+d_{jj})/2$. Other values of σ_{ij} were obtained (26) from vapor-pressure osmometry data.

Table III. Cross osmotic second virial coefficients $B_{ij} \times 10^4$ (ml/g^2) for various polymer and protein pairs in aqueous potassium phosphate buffer. 75 mM, pH 7.0.

	Lysozyme	Dextran T70	PEG3350
Lysozyme	4.1		
Dextran T70	2.3	4.0	
PEG3350	3.35[*]	------	36.3

[*]interpolated between data at 50 mM potassium phosphate and 100 mM potassium phosphate.
SOURCE: Values are from LALLS data tabulated in (26).

Table IV. Salt-Polymer Specific Interaction Coefficients β_{ij} (kg/mol).

	Lysozyme	Dextran T70	PEG3350
KH_2PO_4	1.63	.68	2.67
K_2HPO_4	1.47	.48	1.93

SOURCE: Values taken from (26).

Literature Cited

1. Jaenicke, R. *CIBA Found. Sym.* **1991**, *161*, 206.
2. Chang, B. S.; Randall, C. S. *Cryobiology* **1992**, *29*, 632-656.
3. Arakawa, T.; Timasheff, S. N. *Biochemistry* **1982**, *21*, 6536.
4. Arakawa, T.; Timasheff, S. N. *Biochemistry* **1982**, *21*, 6545.
5. Arakawa, T.; Timasheff, S. N. *Arch. Biochem. Biophys.* **1983**, *224*, 169.
6. Arakawa, T.; Timasheff, S. N. *Biochemistry* **1984**, *23*, 5912.
7. Arakawa, T.; Timasheff, S. N. *Biochemistry* **1984**, *23*, 5924.
8. Arakawa, T.; Timasheff, S. N. *Methods Enz.* **1985**, *114*, 49-77.
9. Arakawa, T.; Timasheff, S. N. *Biochemistry* **1985**, *24*, 6756.
10. Arakawa, T.; Timasheff, S. N. *Biochemistry* **1987**, *26*, 5147.
11. Arakawa, T.; Bhat, R.; Timasheff, S. N. *Biochemistry* **1990**, *29*, 1914.
12. Arakawa, T.; Bhat, R.; Timasheff, S. N. *Biochemistry* **1990**, *29*, 1924.
13. Timasheff, S. N. In *Water in Life*; Somero, G. N., Ed.; Springer-Verlag: Berlin, 1992.
14. Timasheff, S. N. In *Stability of Protein Pharmaceuticals, Part B: In Vivo Pathways of Degradation and Strategies for Protein Stabilization*; Ahern, T. J. and Manning, M. C., Ed.; Plenum Press: New York, 1992.
15. Timasheff, S. N. *Annu. Rev. Biophys. Biomol. Struct.* **1993**, *22*, 67.
16. Carpenter, J. F.; Crowe, J. H. *Cryobiology* **1988**, *25*, 244-255.
17. Heller, M. C.; Carpenter, J. F.; Randolph, T. W. *J. Pharm. Sci.* **1996**, *85*, 1358-1362.
18. Her, L.-M.; Deras, M.; Nail, S. L. *Pharm. Res.* **1995**, *12*, 768-772.
19. King, R. S.; Blanch, H. W.; Prausnitz, J. M. *AIChE J.* **1988**, *34*, 1585-1594.
20. Dong, A.; Prestrelski, S. J.; Allison, S. D.; Carpenter, J. F. *J. Pharm. Sci.* **1995**, *84*, 415-424.
21. Levitt, M.; Greer, J. *J. Mol. Biol.* **1977**, *114*, 181-239.
22. Albertsson, P.-Å. *Partition of Cell Particles and Macromolecules*; 3rd ed.; John Wiley & Sons: New York, 1986.
23. Reinhart, G. D. *J. Biol. Chem.* **1980**, *255*, 10576-10578.
24. Arakawa, T.; Carpenter, J. F.; Kita, A.; Crowe, J. H. *Cryobiology* **1990**, *27*, 401-415.
25. Lee, L. L.-Y.; Lee, J. C. *Biochemistry* **1987**, *26*, 7813-7819.
26. Haynes, C. A., J. Newman, H. W, Blanch, and J. M. Prausnitz. **1993**. *AIChE J.*, *39*, 1539.
27. Haynes, C. A., J. Newman, H. W. Blanch, and J. M. Prausnitz. **1992** Engineering applications of McMillan-Mayer dilute-solution theory, personal communication.
28. Grundke, E.W., and D. Henderson. **1972** *Mol. Phys.*, *24*, 269.
29. Guggenheim, E. A., and R. H. Stokes. **1958**. *Trans. Faraday Soc.*, *54*, 1646.
30. Tanford, C. 1961. Physical Chemistry of Macromolecules. Wiley, New York.
31. Tanford, C., S. A. Swanson, and W. S. Shore. **1955** *JACS*, *77*, 6414.
32. Tanford, C., and M. L. Wagner. **1954**. *JACS*, *76*, 3331.
33. Sakakibara, R., and K. Hamaguchi. **1968**, *J. Biochem.*, *64*, 5, 613.
34. Shiao, D. D. F., R. Lumry, and S. Rajender. **1972**. *Eur. J. Biochem.*, *29*, 377.
35. Horn, D., and C.-C. Heuck. **1983** *J. Biol. Chem.*, *258*, 1665.

36. Haynes, C. A., H. W. Blanch, and J. M. Prausnitz. **1989**, *Fluid Phase Equilibria*, **53**, 463.
37. Tyn, M. T., and T. W. Gusek. **1990** *Biotechnology and Bioengineering*, **35**, 327.
38. Newman, J. **1973**. Electrochemical Systems. Prentice-Hall, Englewood Cliffs, NJ.
39. Shire, S. 1992. Analytical ultracentrifugation and its use in biotechnology, *in* Stability of Protein Pharmaceuticals, Part B: *In Vivo* Pathways of Degradation and Strategies for Protein Stabilization. T. J. Ahern and M. C. Manning, eds., Plenum Press, New York.
40. Rathbone, S. R., C. A. Haynes, H. W. Blanch, and J. M. Prausnitz. **1990**. *Macromolecules*, **23**(17), 3948
41. Kratochvil, P., J. Vorlicek, D. Strakova, and Z. Tuzar. 1975. Light scattering investigations of interaction between polymers in dilute solution, *J. Polymer Sci., Polymer Phys. Ed.*, **13**, 2321.

Chapter 6

Phase Separation and Crystallization of Components in Frozen Solutions: Effect of Molecular Compatibility Between Solutes

Ken-ichi Izutsu, Sumie Yoshioka, and Shigeo Kojima

National Institute of Health Sciences, 11–18–1 Kamiyoga, Setagaya-158, Tokyo, Japan

The physical state of components in a frozen solution and a freeze-dried solid is an important factor that determine protein stability. During the freezing of aqueous solutions containing various combinations of polyols, a fraction of the liquid does not freeze, but remains as a concentrated solution among ice crystals. This unfrozen fraction may form an amorphous mixture phase or undergo further phase separation into liquid or solid phases. Frozen solutions containing some combinations of non-crystallizing polymers (e.g., polyvinylpyrrolidone (PVP) and dextran) show two softening temperatures (T_s) close to their individual pure-component T_s. Although some sugars and polymers (e.g., glucose, sucrose, PVP) inhibit the crystallization of poly(ethylene glycol)(PEG) in frozen solutions by forming a practically stable amorphous mixture phases, others (e.g., melibiose, dextran) have little effect on PEG crystallization. A PEG cloud point depression study showed that mono- and disaccharides that are relatively less miscible with PEG in solution inhibit PEG crystallization to a lesser degree. These results suggests that phase separation in frozen solutions occurs through a mechanism similar to those which cause separation of aqueous two-phase systems at room temperature. Possible implications of the phase separation phenomena to the stability of protein formulations are discussed.

The development of many recombinant proteins has brought about an increasing need for formulations that stabilize proteins. Although freeze-drying is often the process of choice to achieve long term storage stability, many proteins lose their biological activity during the process, mainly by physical inactivation mechanisms such as denaturation and aggregation. Appropriate stabilizing additives for protein pharmaceutical products and the mechanisms responsible for their action have been studied extensively (*1-13*).

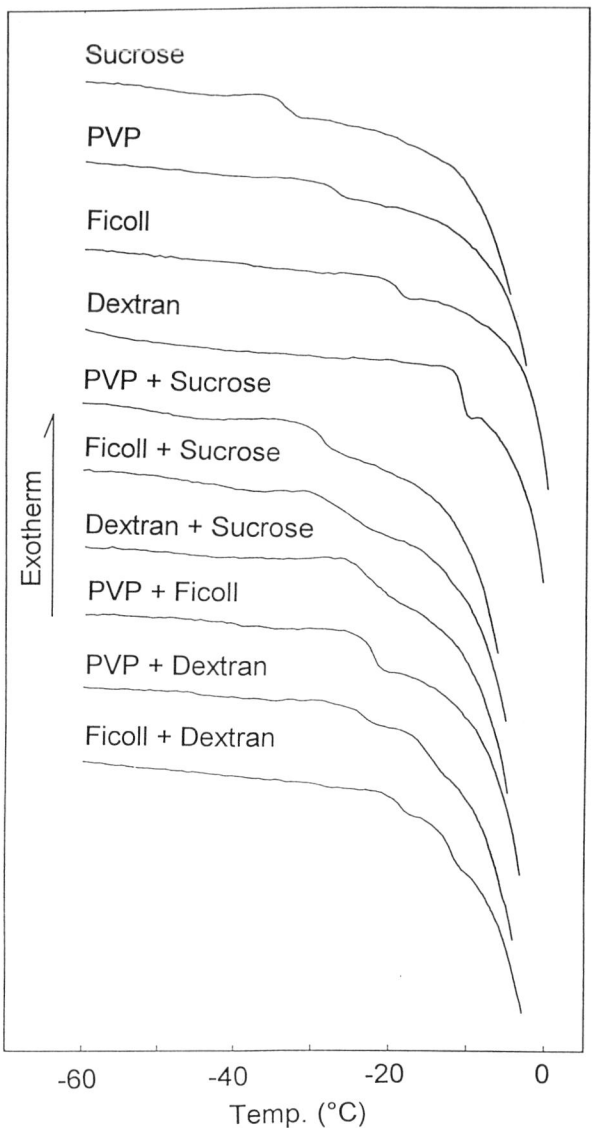

Figure 1. DSC profiles of frozen solutions containing various sugars, polymers and their mixtures (50 mg/ml for each solute). Reproduced with permission from reference 4. Copyright 1996 Plenum Publishing Corporation.

The freezing and freeze-drying of dilute aqueous solutions are complex processes involving various physical phenomena (1,2). Ice usually crystallizes first as the solution cools. The physical characteristics of the residual concentrated solution phase depend on the kind of solutes, their concentration and freezing method. Some excipients (NaCl, mannitol, PEG) readily crystallize as binary ice/solute crystals. Other solutes (glycerol, sucrose, dextran and most other carbohydrates) form kinetically stable amorphous phase without crystallization. The crystallinity of each component depends on the complex interplay of molecular interactions during freezing and freeze-drying of multi-component pharmaceutical formulations. The physical states of the "active" and other components are important factors in determining the quality of pharmaceutical products, such as appearance, solubility and stability (3,14). Understanding the physical phenomena that happen during freezing and freeze-drying and controlling them are necessary for rational design of protein formulations.

A supercooled solutions formed among the ice crystals have usually been treated theoretically as a single uniform amorphous phase. However, recent studies show the presence of multiple amorphous phases due to phase separation in frozen solutions containing more than one solute (4,15). Although the mechanism of the phase separation in frozen solutions has not been well understood, this phenomenon could affect the quality of freeze-dried formulations because it alters the molecular interactions between components. In this chapter we describe two types of aqueous polymer solution models which indicate phase separation in frozen solutions. One of them cause formation of multiple amorphous phases, the other shows crystallization of a solute that depends on its miscibility with other components in the solution. Possible effect of the phenomenon on to the stability of freeze-dried formulations is discussed.

Phase Separation into Multiple Amorphous Phases

Thermal analysis by differential scanning calorimetry (DSC) of frozen solutions containing a non-crystallizing solutes usually shows two transition regions (15-17). They are the "real" glass transition (T_g) and softening temperature of the amorphous phase (T_s, 18). The T_s has been called the glass transition temperature of maximally freeze-concentrated solution (T_g'). Her et al. reported the existence of multiple T_ss (termed T_g' in the paper, 15) in frozen solutions containing both polyvinylpyrrolidone (PVP) and phosphate buffer components. Addition of 2.0 % sodium phosphate buffer to 10% PVP caused a second T_s region to appear, while addition of other salts showed single T_s. The two T_s regions suggest the presence of multiple amorphous phases in the freeze-concentrate (15).

The presence of multiple amorphous phases was also observed in frozen solutions containing two non-crystallizing polymers. Figure 1 shows DSC scans of the frozen solutions containing sucrose and/or non-crystallizing polymers. All the solutions subjected to thermal analysis were in a single phase at room temperature. The amorphous phase in frozen solutions exhibits a T_s that is characteristic of the solutes (1,19). The T_ss for sucrose, PVP, Ficoll and dextran solutions are consistent with those in other reports (*1*). Frozen solutions that contain both sucrose and polymers, or PVP and Ficoll, show a single T_s region which lies between the T_s values for the individual pure components, indicating formation of a single amorphous phase in frozen solution. On the other hand, frozen solutions that contain dextran and Ficoll, or dextran and PVP, show two T_s regions that are close to their respective pure component solutions. The "real" glass transition of the frozen solution (T_g) was not detectable.

The presence of two amorphous phases in a frozen solution was observed in combinations of polymers (or polymer and salt) that form aqueous two-phase systems. Aqueous solutions containing high concentrations of PVP and phosphate buffer separate into two liquid phases (*20*). Concentrated solutions containing combinations of dextran and Ficoll, or dextran and PVP also form aqueous two-phase systems at room temperature (*20,21*), suggesting that the phase separation in frozen solutions occur through a mechanism similar to that causes aqueous two phase separation in room temperature solutions. Aqueous solutions containing more than a certain concentration of two chemically different polymers (e.g., PVP and dextran), one polymer and a salt (e.g., PVP and potassium phosphate) or one polymer at high temperature (e.g., PEG) separate into two immisible phases (*20-24*). Each phase is rich in one of the polymers (or salt). Various combinations of polymers (e.g., PEG, PVP, dextran, polyvinyl alcohol (PVA), Ficoll, methylcellulose, casein) capable of phase separation in water have been reported. The two T_s values in frozen solutions containing both PVP and dextran, or Ficoll and dextran were not identical to those of solutions made of their pure components. This is to be expected, because both of the separated phases should be rich in one polymer and contain a small amount of the other polymer (*22-24*).

The phase separation (two amorphous phases) in frozen solution was observed at an overall concentration much lower than that required to cause phase separation at room temperature. Although the concentration of each polymer required for phase separation in an aqueous solution usually decrease with decreasing temperature (*22,24-28*), the temperature dependence of the phase boundary is insufficient to explain the phase separation observed in frozen solutions at a low overall concentration of polymers. It is plausible that freezing-induced increases in solute concentration can lead to such phase separation during the freezing process. The separation between non-crystallizing solute phases in frozen solution should result in microscopically heterogeneous amorphous solid after freeze-drying.

Phase Separation and Crystallization of Components in Frozen Solutions

Solute crystallization in frozen solution depends on their nature and concentration of the solute, the presence of other formulation components, and the details of freezing process (*5,29-32*). Although most of the carbohydrates used for pharmaceutical formulations remain amorphous in frozen solutions, some (e.g., mannitol, inositol, PEG) crystallize (*6,34-36*). PEG crystallizes in frozen solutions as PEG/water complexes (containing two or three water molecules on the average per molecule of PEG) (*34*). The crystallization of solutes in frozen solution is inhibited by the addition of non-crystallizing sugars and polymers (termed "third components" of the solution)(*33*). Since aqueous solutions containing PEG and various polymers form aqueous two-phase systems, the relationship between the ability of various third components to inhibit PEG crystallization in frozen solution and their miscibility in solution was studied. Thermal analysis of a frozen aqueous solution of PEG 3000 shows an exothermic peak at around -50 °C and an endothermic peak at around -15 °C, which are formation (crystallization) and melting of PEG/water complexes, respectively (*34*). Table I shows the effects of various low-molecular-weight saccharides and polymers on the crystallization of 20 mg/ml PEG 3000, based on the enthalpy of the melting endotherm. The crystallization of PEG 3000 is completely inhibited by addition of 20 mg/ml glucose, fructose, sucrose, palatinose and maltose. The inhibition of crystallization by third components results in the absence of PEG crystals in freeze-dried cakes (*7,20*). In contrast, the crystallization of the PEG is not fully inhibited by up to 200 mg/ml of melibiose, Ficoll and dextran. Molecular mobility of water and solute molecules studied by pulsed NMR showed formation of an amorphous PEG/sucrose/water mixture phase in the frozen solution, while frozen

Table I. Effect of Third Components on Crystallinity of PEG 3000 in Frozen Solutions

	Crystallinity of PEG/water Systems(%)*							
	Third Components (mg/ml)							
	5	10	15	20	50	100	150	200
Glucose	62.6	43.5	1.5	0.0	0.0	0.0	0.0	0.0
Fructose	78.0	45.9	5.2	0.0	0.0	0.0	0.0	0.0
Galactose	79.6	59.5	6.4	5.8	9.6	2.8	0.0	0.0
Sucrose	71.7	42.2	0.7	0.0	0.0	0.0	0.0	0.0
Palatinose	72.9	60.5	12.3	0.0	0.0	0.0	0.0	0.0
Maltose	80.4	73.0	14.2	0.3	0.0	0.0	0.0	0.0
Cellobiose	80.6	63.4	4.0	0.7	0.0	0.0	0.0	0.0
Trehalose	82.2	82.1	55.0	14.5	0.0	0.0	0.0	0.0
Lactose	87.6	60.5	41.5	74.6	65.1	47.3	2.4	0.0
Melibiose	89.0	89.5	91.6	90.7	85.5	76.0	57.1	57.5
Raffinose	90.4	83.2	34.8	4.8	0.9	0.0	0.0	0.0
PVP10000	86.4	80.3	75.5	75.5	2.7	0.0	0.0	0.0
Ficoll	98.5	96.2	94.8	90.1	84.5	67.3	53.9	49.7
Dextran	94.7	92.7	98.6	94.5	92.5	93.1	85.1	87.2

*Crystallinity of a PEG/water system is described as the ratio (%) to that of a 20 mg/ml PEG3000 solution. Reproduced with permission from reference 4.

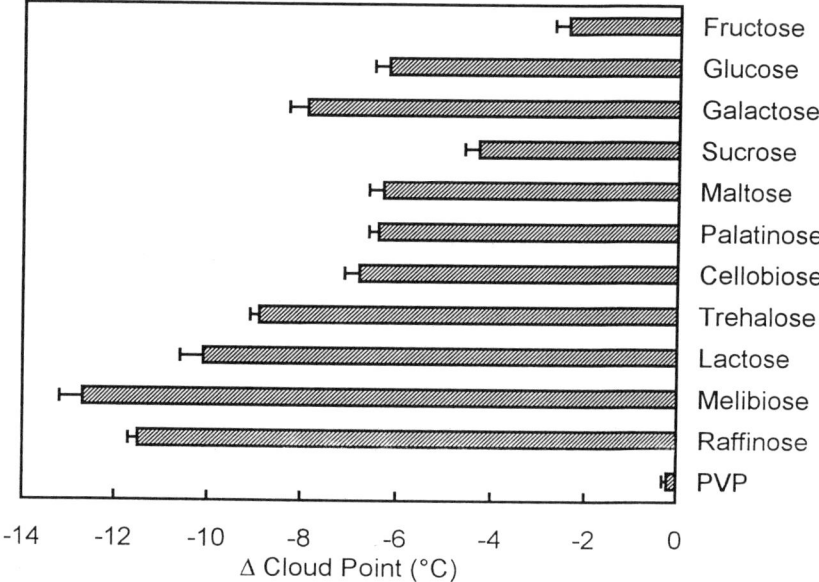

Figure 2. Effect of third components (10% w/w) on the cloud point of 10% (w/w) PEG 20,000. Values are expressed as means and SD of three experiments. Cloud point of the 10% (w/w) PEG 20,000 solution was observed at 115.4 °C. Reproduced with permission from reference 4. Copyright 1996 Plenum Publishing Corporation.

solutions containing PEG and dextran showed independent behavior of each component (*4*). The variation was much larger than that of the concentration required to inhibit crystallization of NaCl/water systems (*37*). Among monosaccharides, a greater concentration of galactose was required to inhibit the crystallization of PEG 3000 compared to that required for fructose and glucose. The order of the effect among disaccharides was sucrose > maltose, palatinose, cellobiose > trehalose > lactose > melibiose. Addition of 100 mg/ml PVP (average MW: 10,000) completely inhibited the crystallization of PEG 3000, but dextran (average MW: 11,000) showed little effect. No apparent relationship was observed between ability of these solutes to inhibit the crystallization of PEG and their T_gs (*17,19,38*).

Effect of Molecular Interaction in Solution on Phase Separation in Frozen Solution

An aqueous solution of PEG has a lower critical solution temperature (LCST), in which liquids that are completely miscible at a lower temperature form a two-phase system at higher temperatures (*39-41*). The temperature at which this phenomenon occurs is also called the cloud point. Many third components, including low-molecular-weight saccharides, decrease the cloud point of PEG (*39,40*). Since the low-molecular-weight saccharides are segments of polysaccharides that form aqueous two-phase system with PEG, the mechanism of the PEG cloud point lowering by these saccharides is fundamentally identical to formation of an aqueous two-phase system composed of PEG and dextran (*39*). The cause of the phase separation and cloud point lowering are described by repulsive interaction between polymer segments (in Flory-Huggins theory, *20,26,39*) and/or incompatibility of hydrated water structure (*21,22*). Cloud point data thus provide information on PEG-excipient interactions that are not repulsive enough to form aqueous two-phase systems at low and moderate temperatures (*39*).

Figure 2 shows the effects of third components (10%, w/w) on the cloud point of a 10% (w/w) solution of PEG 20,000. The PEG 20,000 solution without third components shows a cloud point at 115.4 °C. Among the monosaccharides studied, galactose shows the greatest effect in lowering the cloud point, whereas fructose showed the smallest effect. The effect of disaccharides in depressing the cloud point occurs in the order sucrose < maltose, palatinose, cellobiose < trehalose < lactose < melibiose. Significant differences in the effects of polymers were observed. Addition of 10% (w/w) dextran or Ficoll results in phase separation at room temperature, whereas PVP has little effect on the cloud point of PEG 20,000. A larger decrease in PEG cloud point by the third component corresponds to less miscibility, or a more repulsive interaction, between the pair. The results in Figure 2 indicate that interactions between PEG and some saccharides (melibiose, lactose, trehalose) are more repulsive than that of PEG and sucrose.

The results in Table I and Figure 2 indicate that the sugars that are less miscible with PEG in solution have smaller inhibitory effect on PEG crystallization in frozen solutions. The order of mono- and disaccharides that inhibit the crystallization of PEG in frozen solution was just opposite to that causing cloud point lowering. As was observed in aqueous solutions of non-crystallizing polymers such as PVP and dextran, the solution containing PEG and dextran or Ficoll should phase separate during freeze-concentration. PEG should crystallize in the separated phase in the frozen solution. Crystallization and melting of PEG is not affected by dextran, suggesting that phase separation occurs during the cooling process.

The effect of low-molecular-weight saccharides on PEG crystallization depends on their interaction strengths. Saccharides such as glucose, fructose and sucrose provide a sufficient hydrogen bonding network (42) in amorphous PEG/sugar/water systems to keep PEG from crystallizing. Increase in PEG crystallization temperature in solutions containing PEG and lactose indicates that the amorphous phases apparently do not separate during the freezing process. It suggests that increased repulsive interactions between components (e.g., PEG-lactose) result in failure of the hydrogen bonding network and subsequent crystallization of PEG, thus raising the saccharide/PEG ratio required to inhibit crystallization of PEG.

The miscibility between low-molecular-weight saccharides and PEG depends on the structure of saccharides. The disaccharides employed for the study can be roughly divided into three groups. Disaccharides composed of glucose and galactose (lactose, melibiose) are less compatible with PEG in solution than those with glucose and fructose (sucrose, palatinose). Disaccharides composed of two glucose moieties (maltose, cellobiose, trehalose) fall in the middle of the groups above. Structural difference between these saccharides is a relative position of OH(4) and OH(2) that determines their hydration state (39,43,44). Galactose and disaccharides containing galactose subunits have relatively poor compatibility with the three-dimensional hydrogen-bonded structure of water relative to that of fructose, glucose, and their respective disaccharides (43). The difference in the compatibility between the saccharide and PEG should depend on different hydrophilicity of saccharides (39) and/or their different effect on water structure (21,22).

Possible Effect of Phase Separation on Protein Formulations

Although the phase separation in frozen is still not well understood, it should be one of the important factors that determine quality of freeze-dried protein pharmaceuticals. Formation of aqueous two-phase system in various combinations of polymers (or polymer and salt) often used in pharmaceutical formulations have been reported (22,45). The combinations include that of proteins and other polymers (e.g., casein and carboxymethyl cellulose, globulin and dextran). It is plausible that the phase separation in frozen solution occurs between protein and additive or between different additives in multi-component protein formulations. The phase separation in frozen solution should result in microscopically heterogeneous freeze-dried solid.

Various additives, including low-molecular saccharides and polymers, are added to stabilize proteins in formulations. The phase separation between protein and additives should remove molecular interaction between them needed to maintain protein structure in dried state (3,6-10,46,47). For example, although PEG is potent stabilizer in freeze-thawing of proteins, it loses this effect in freeze-drying due to crystallization

(6). If PEG remains amorphous by co-lyophilizing with third components (e.g. sucrose), it provides a significant protein-stabilizing effect in freeze-drying (7). In addition, alternation of various physico-chemical properties of freeze-dried cake such as T_g and crystallinity by phase separation should be another factor that affects quality of freeze-dried formulations. It is clear that extensive studies on phase separation in frozen solution and controlling of a the phenomena are required to develop reliable freeze-dried protein formulations.

Acknowledgments
Part of the work is a collaboration with Dr. J. F. Carpenter of the School of Pharmacy, University of Colorado Health Sciences Center and Dr. T. W. Randolph of Department of Chemical Engineering, University of Colorado. We gratefully acknowledge support from National Science Foundation Grant BES9505301 and Japan Health Sciences Foundation.

Literature Cited

1. Franks, F. *Jpn. J. Freezing Drying* **1992**, 38, 5-16.
2. Franks, F. *Process Biochem.* **1989**, 24, S3-S8.
3. Arakawa, T.; Kita, Y.; Carpenter, J.F. *Pharm. Res.* **1991**, 8, 285-291.
4. Izutsu, K.; Yoshioka, S.; Kojima, S.; Carpenter, J.F.; Randolph, T.W. *Pharm. Res.* **1996**, 13, 1393-1400.
5. Pikal, M.J. *Biopharm* **1990**, 3, 26-29.
6. Carpenter, J.F.; Prestrelski, S.J.; Arakawa, T. *Arch. Biochem. Biophys.* **1993**, 303, 456-464.
7. Izutsu, K.; Yoshioka, S.; Kojima, S. *Pharm. Res.* **1995**, 12, 838-843.
8. Townsend, M.W.; DeLuca, P.P. *J. Parenteral Sci. Tech.* **1988**, 42, 190-199.
9. Carpenter, J.F.; Prestrelski, S.J.; Anchordoguy, T.J.; Arakawa, T. In *Formulation and Delivery of Protein and Peptides*, Cleland, J.L.; Langer, R. Eds. ACS Symposium Series 567, Washington D.C., 1994, pp. 134-147.
10. Dong, A.; Prestrelski, S.J.; Allison, S.D.; Carpenter, J.F. *J. Pharm Sci.*, **1995**, 84, 415-424.
11. Arakawa, T.; Timasheff, S.N. *Biochemistry* **1982**, 21, 6536-6544.
12. MacKenzie, A.P. *Dev. Biol. Stand.* **1976**, 36, 51-67.
13. Manning, M.C.; Patel, K.; Bocahardt, R.T. *Pharm. Res.* **1989**, 6, 903-918.
14. Pikal, M.J.; Lukes, A.L.; Lang, J.E. *J. Pharm. Sci.* **1977**, 66, 1312-1316.
15. Her, L.M.; Deras, M.; Nail, S.L. *Pharm. Res.* **1995**, 12, 768-772.
16. Rasmussen, D.; Luyet, B. *Biodynamica* **1969**, 10, 319-331.
17. Chang, B.S.; Randall, C.S. *Cryobiology* **1992**, 29, 632-656.
18. Shalaev, E.Y.; Franks, F. *J. Chem. Soc., Faraday Trans.* **1995**, 91, 1511-1517.
19. Suzuki, E.; Nagashima, N. *Bull. Chem. Soc. Jpn.* **1982**, 55, 2730-2733.
20. Walter, H.; Johansson, G.; Brooks, D.E. *Anal. Biochem.* **1991**, 197, 1-18.
21. Zaslavsky, B.Y.; Bagirov, T.O.; Borovskaya, A.A.; Gulaeva, N.D.; Miheeva, L.H.; Mahmudov, A.U.; Rodnikova, M.N. *Polymer* **1989**, 30, 2104-2111.

22. Zaslavsky, B.Y. *Aqueous Two Phase Partitioning: Physical Chemistry and Bioanalytical Applications*, Marcel Dekker, New York, 1994.
23. Walter, H.; Johansson, G. *Anal. Biochem.* **1986**, 155, 215-242.
24. Albertsson, P.-A. *Partition of Cell Particles and Macromolecules*, 3rd. ed., Wiley-Interscience, New York, 1986.
25. Sjöberg, A.; Karlström, G. *Macromolecules* **1989**, 22, 1325-1330.
26. Suzuki, T.; Franks, F. *J. Chem. Soc., Faraday Trans.* **1993**, 89, 3283-3288.
27. Diamond, A.D.; Hsu, J.T. *Biotechnol. Techniques* **1989**, 3, 119-124.
28. Forciniti, D.; Hall, C.K.; Kula, M.R. *Fluid Phase Equilib.* **1991**, 61, 243-263.
29. DeLuca, P.; Lachman, L. *J. Pharm. Sci.* **1965**, 54, 1411-1415.
30. Patel, R.M.; Hurwitz, A. *J. Pharm. Sci.* **1972**, 61, 1806-1810.
31. Akers, M.J.; Milton, N.; Bryn, S.R.; Nail, S.L. *Pharm. Res.* **1995**, 12, 1457-1461.
32. Murase, N.; Franks, F. *Biophys. Chem.* **1989**, 34, 293-300.
33. Jochem, M.; Körber, C.H. *Cryobiology* **1987**, 24, 513-536.
34. Bogdanov, B.; Mihailov, M. *J. Polymer Sci.: Polymer Phys. Ed.* **1985**, 23, 2149-2158.
35. Gatlin, L.; DeLuca, P. *J. Parenter. Drug Assocn.* **1980**, 34, 398-408.
36. Williams, N.A.; Lee, A.; Polli, G.P.; Jennings, P.A. *J. Parenteral Sci. Tech.* **1986**, 40, 135-140.
37. Izutsu, K.; Yoshioka, S.; Kojima, *Chem. Pharm. Bull.* **1995**, 43, 1804-1806.
38. Levine, H.; L. Slade, L. *J. Chem. Soc., Faraday Trans. 1* **1988**, 84, 2619-2633.
39. Sjöberg, A.; Karlstrom, G; Tjerneld, F. *Macromolecules* **1989**, 22, 4512-4516.
40. Schott, H.; Royce, A.E. *J. Pharm. Sci.* **1984**, 73, 793-799.
41. Karlström, G. *J. Phys. Chem.* **1985**, 89, 4962-4964.
42. Oguchi, T.; Terada, K.; Yamamoto K.; Nakai,Y. *Chem. Pharm. Bull.* **1989**, 37, 1881-1885.
43. Galema, S.A.; Høiland, H. *J. Phys. Chem.* **1991**, 95, 5321-5326.
44. Franks, F. *Cryobiology* **1983**, 20, 335-345.
45. Tolstoguzov, V.B. *Food Hydrocolloids*, **1988**, 2, 195-207.
46. Nema, S.; Avis, K.E. *Parenteral Sci. Tech.* **1993**, 47, 76-83.
47. Hellman, K.; Miller, D.S.; Cammack, K.K. *Biochem. Biophys. Acta.* **1983**, 749, 133-142.

Chapter 7

In Situ Formation of Polymer Matrices for Localized Drug Delivery

Jennifer L. West

Institute for Biosciences and Bioengineering, Rice University, Mail Stop #144, Houston, TX 77251–1892

Polymer matrices can be formed within the body in direct contact with cells and tissues by a variety of techniques, both physicochemical and covalent in nature, including thermal gelation, pH-sensitive gelation, the use of thixotropic materials, precipitation from certain organic solutions, photopolymerization, and chemical cross-linking. These types of materials can be injected as liquids and converted to a gel or solid in situ. Many of these polymer systems are appropriate for delivery of peptides and proteins.

Advances in biotechnology have resulted in the development of a number of new biological pharmaceuticals, such as proteins, peptides, and oligonucleotides. Systemic delivery is often ineffective for these drugs, due to undesirable side effects, short half-lives, and drug metabolism. Relatively hydrophilic polymer matrices are usually most suitable for local, sustained delivery of proteins and other biological pharmaceuticals. In addition, continuing trends towards laparascopic and catheter-based surgical procedures have created a significant need for polymeric drug delivery systems that can be applied to tissues in liquid precursor form and converted to a gel or solid state in situ to provide sustained and localized delivery of therapeutic agents at the site of surgical intervention. Any drug that is dissolved in the liquid precursor solution is then homogeneously dispersed in the polymer matrix and subsequently released via diffusion or polymer degradation. In situ formation of solid polymeric drug delivery systems could also be beneficial for opthalmic, buccal, vaginal, rectal, and subcutaneous delivery routes. There currently exist relatively few polymeric materials that can be formed in situ from injectable liquids. Methods for the formation of polymer matrices can be generally classified into two categories, physicochemical and covalent cross-linking. Substantial limitations exist for both types of cross-linking; physicochemically cross-linked materials usually have mild processing conditions but poor mechanical properties and relatively short durations of release of bioactive agents. Covalent cross-linking generally provides stronger and more stable polymer matrices but can have significant cytotoxicity due to the initiators and monomers utilized as well as the heat produced during the reaction. Several modes of in situ polymer matrix formation, both physicochemical and covalent in nature, are discussed below, including thermal gelation, pH-sensitive polymers, shear-thinning polymers, precipitation of polymers from biocompatible organic solutions, photopolymerization, and chemical cross-linking. Several of these systems have been investigated for delivery of proteins and other bioactive agents in vivo.

Thermal Gelation.

Temperature-sensitive polymer systems have been investigated for local delivery of proteins and oligonucleotides in vivo. The most extensively utilized materials of this type are tri-block copolymers of polyethylene glycol and polypropylene glycol, termed Pluronics or Poloxamers, that form micelles in aqueous solutions. At low temperatures (below 10° C), these materials are liquids, and at physiological temperatures, they are converted to hydrogels due to aggregation of the micelle structures (1). Thus, chilled polymer solutions can be injected, with hydrogel formation occurring in situ as the solution is warmed to body temperature: this phase transition generally occurs over a period of several seconds. Pluronic F-127 (Poloxamer 407) has been evaluated for drug release in several studies in vivo. Pluronic F-127 hydrogels have been formed around the outside of an artery after balloon-induced injury in the rat for localized delivery of anti-sense c-myb oligonucleotides: intimal thickening, a wound healing complication after vascular injury, was significantly reduced by this treatment, and the activity of the oligonucleotide was effectively localized to the region of hydrogel application (2). Pluronic F-127 hydrogels were used for sustained intraperitoneal administration of the chemotherapeutic agent mitomycin C in order to enhance its effects against a Sarcoma-180 ascites tumor in mice; mitomycin C was more effective at the same total dose when release from the Pluronic gel than when given as an intraperitoneal injection (3). Enzymatic activity of a model protein, urease, incorporated into pluronic F-127 hydrogels was evaluated after intraperitoneal injection and in situ hydrogel formation in the rat. The biological activity of the entrapped urease was maintained, and hyperammonemia was extended approximately three-fold compared with rats receiving administration of the same dose of urease in phosphate buffered saline (4). A thermally gelling poloxamine solution (Synperonic T908) has also been evaluated for use as a drug delivery vehicle. These gels have been used successfully for rectal administration of indomethacin in rabbits (5). Thermal gelation of polymer matrices is an attractive methodology for use in vivo because the conditions for matrix formation are generally mild and most of the materials investigated have had relatively good biocompatibility properties. However, these materials are sometimes difficult to handle due to premature gelation, and the mechanical properties are inadequate and release rates too rapid for some applications.

pH-Sensitive Polymers. In situ formation of polymer matrices may be accomplished by using materials that are soluble at either moderately low or moderately high pH but insoluble at physiological pH; either anionic or cationic moieties within the polymer may be used to impart this behavior to a material. Such polymers can be applied to tissues as a liquid with subsequent gelation occurring as the pH approaches physiological values. Similar cross-linked polymer systems, where the degree of matrix swelling depends on pH, have been investigated for pH-responsive drug delivery. Solutions of polyacrylic acid (PAA) can be applied to tissues as a liquid and precipitate as they reach physiological pH; PAA solutions are low viscosity liquids at pH 4 and transform into stiff gels at pH 7.4. In situ gelation of PAA solutions has been evaluated for opthalmic drug delivery (6). These materials are easy to apply, but exposure to very acidic conditions may cause local tissue damage. The addition of viscous polymers, such as hydroxypropyl methylcellulose, may improve the mechanical properties of these materials, allowing lower amounts of PAA and solutions with a milder initial pH to be employed (7).

Shear-thinning Polymers. Thixotropic polymers, materials that exhibit reduced viscosity under shear, have been used in several tissue-contacting applications. These types of materials form polymer networks via chain entanglement in the absence of shear, resulting in the formation of gels or highly viscous liquids, but upon application of shear, the polymer chains become aligned, causing the chain entanglements to disengage and allowing the material to flow. Thus, thixotropic polymers may be injected under shear with matrix formation occurring in contact with tissue upon cessation of flow. Hyaluronic acid derivatives exhibiting this type of behavior have been utilized as barriers to prevent post-surgical adhesion formation in rat and rabbit models (8). These materials exhibited rapid gelation and good retention at the site of application. Sodium hyaluronate solutions exhibit thixotropic behavior, forming highly viscous liquids in the absence of shear, and have been evaluated for use as artificial tears. Sodium hyaluronate solutions have been shown to have a longer ocular residence time than polymer solutions with Newtonian rheologic properties (9). Thixotropic materials do not appear to have been utilized for localized delivery of proteins or peptides to date, but would conceivably be an effective means for sustained local delivery.

Precipitation from Organic Solutions. Solid polymer matrices may be formed in situ by injecting a solution of a water-insoluble polymer in a reasonably biocompatible solvent. As the solvent diffuses into tissues, the polymer precipitates and solidifies. Solutions of poly(DL-lactide) in N-methyl-2-pyrrolidone can be injected to form solid polymers in situ that are appropriate for localized and sustained protein release (10). This type of polymer system has been utilized to deliver growth factors to stimulate bone healing in a rabbit model (11). Protein release from these systems is first-order, and the matrix formation method appears to have little impact on protein activity (12). The lack of cytotoxicity and the minimal impact on protein activity may be specific to this particular polymer/solvent system.

Photopolymerization. Several polymer systems have been developed wherein aqueous solutions of polymer precursors containing reactive moieties are applied to tissues and subsequently covalently cross-linked to form hydrogels that adhere and conform to the underlying tissues. Photopolymerization offers some benefits over many other modes of chemical cross-linking in that the reaction is generally very rapid and cross-linking can be spatially constrained by selective exposure to light. Furthermore, many of the systems that have been developed involve photopolymerization of macromolecular precursors, thus avoiding issues with monomer toxicity and reducing heat formation.

Aqueous solutions of acrylated block copolymers of polyethylene glycol and oligomers of α-hydroxy acids can be photopolymerized in contact with cells and tissues over a period of several seconds to form biocompatible hydrogel materials that are suitable for delivery of proteins, peptides, and oligonucleotides (13). Drugs are dissolved in the aqueous precursor solution and entrapped within the hydrogel matrix following photopolymerization induced by low intensity long wavelength ultraviolet light with 2,2-dimethoxy-2-phenyl acetophenone as a photoinitiator. Hydrogel formation is complete within 30 seconds. These hydrogels have been utilized for the localized delivery of fibrinolytic agents, such as tissue plasminogen activator (tPA) and urokinase plasminogen activator, for the prevention of post-surgical adhesion formation after uterine horn devascularization in the rat (14). tPA activity was found to remain substantially unchanged following photopolymerization and release into aqueous media in vitro (15). Additionally, these types of polyethylene glycol-based polymers can be used to synthesize thin hydrogel coatings (< 50 µm) on tissues via an interfacial photopolymerization technique wherein a non-toxic photoinitiator is adsorbed onto the tissue surface, the tissue is coated with the precursor solution, and the liquid precursor is converted to a hydrogel only at the tissue surface upon exposure to an appropriate light source. This process can be completed using catheter-based devices and polymerization occurs over a period of several seconds. Thin hydrogel barriers have been formed on the luminal surface of arteries following vascular injury to prevent thrombosis (16). Such barriers may also be utilized for localized delivery of therapeutic agents, such as heparin, to the arterial wall (17).

Several other photopolymerizable polymers have been developed that can be cross-linked in contact with tissue, though drug release from these materials has not yet been reported. Branched polyethylene glycols with pendant cinnamylidene acetyl moieties can be cross-linked by exposure to ultraviolet light (300 nm) in the absence of photoinitiators (18). Hydrogel formation occurred in this system within approximately 5 min. This system is particularly attractive, as most initiators are somewhat toxic. In addition, solutions of cinnamated hyaluronic acid or chondroitin sulfate can be photopolymerized to form coatings that conform and adhere to the underlying tissue; these coatings have been found to prevent post-surgical adhesion formation after peritoneal injury in the rat (19).

Chemical Cross-linking. Covalent cross-linking of polymeric materials may be carried out in situ using methods other than photopolymerization. These types of systems for in situ polymer matrix formation will generally be "two-syringe" methods, wherein the polymer precursor and the initiator are injected separately with mixing occurring in contact with tissues. As described above, the concerns of monomer and initiator toxicity and heat production must be addressed in the development of such systems, and organic solvents should be avoided due to issues with biocompatibility and protein activity. Thus, ideally, only aqueous solutions of macromolecular precursors should be utilized. Several systems based on in situ chemical cross-linking of water soluble, macromolecular precursors have been developed, though drug release from these materials does not appear to have been evaluated to date.

Hydrogels can be formed in situ by mixing a concentrated solution of serum albumin with an aqueous solution of polyethylene glycol disuccinimidylsuccinate (20). These materials have been evaluated in rabbits as tissue sealants: the hydrogels form within a few seconds of mixing, are bioresorbable, and do not cause marked tissue damage. Hydrogels may also be formed in situ by mixing gelatin with water-soluble carbodiimides (21). These hydrogels have been used as tissue adhesives in mice and were found to induce very little inflammation or cytotoxicity. Polylactide diisocyanate materials may also be cross-linked in situ; in this case, water causes the polymerization reaction, so matrix formation occurs upon mixing with biological fluids rather than with a second injected substance (22). Hydrophilic units can be added to the polymer chain to increase the rate of water absorption and thus the rate of cross-linking. Chemically cross-linked polymer matrices offer better mechanical properties and more sustained drug release than physicochemical polymer matrices, but possible cytotoxicity and preservation of the activity of entrapped bioactive agents will have to be addressed.

Conclusions. In situ formation of polymer matrices offers a number of advantages over conventional implantation of pre-formed polymer matrices for localized drug delivery: laprascopic and catheter-based procedures can be used for polymer implantation, the shape of the polymer devices can conform precisely to that of the underlying tissue, and in most cases the polymer matrices adhere firmly to the underlying tissue, allowing for effective localization of drug release. Most of these methods for polymer matrix formation utilize aqueous, macromolecular precursor solutions and mild processing conditions, requisite for biocompatibility of the formation process, and are thus suitable vehicles for delivery of peptides and proteins. Several of these materials, such as temperature-sensitive Pluronic polymers

(2) and photopolymerized polyethylene glycol-based hydrogels (*14*), have been used successfully for localized, sustained delivery of proteins in animal models. In addition, because proteins or other hydrophilic, bioactive agents can be dissolved in the aqueous precursor solutions, the resultant polymer matrices contain homogeneously dispersed drug molecules. In most solid polymer formulations, homogeneous drug dispersion is difficult to achieve and release profiles are thus somewhat unpredictable.

This review has covered a number of methods for the formation of polymer matrices in contact with tissue. These materials can be either covalently or physicochemically cross-linked. In all likelihood, many other materials that can be formed in situ will be developed for delivery of peptides, proteins, and oligonucleotides. Efforts should focus on the synthesis of materials that are highly biocompatible, have appropriate mechanical properties, provide sustained release, and do not have adverse effects upon activity of entrapped bioactive species.

References

1. Leach, R. E.; Henry, R. L. *Am. J. Obstet. Gynecol.* **1990**, *162*, 1317.
2. Simons, M.; Edelman, E. R.; DeKeyser, J. L.; Langer, R.; Rosenberg, R. D. *Nature* **1992**, *359*, 67.
3. Miyazaki, S.; Ohkawa, Y.; Takada, M.; Attwood, D. *Chem. Pharm. Bull.* **1992**, *40*, 2224.
4. Pec, E. A.; Wout, Z. G.; Johnston, T. P. *J. Pharm. Sci.* **1992**, *81*, 626.
5. Miyazaki, S.; Oda, M.; Takada, M., Attwood, D. *Biol. Pharm. Bull.* **1995**, *18*, 1151.
6. Kumar, S.; Himmelstein, K. J. *J. Pharm. Sci.* **1995**, *84*, 344.
7. Kumar, S.; Haglund, B. O.; Himmelstein, K. J. *J. Ocul. Pharmacol.* **1994**, *10*, 47.
8. Burns, J. W.; Skinner, K.; Lu, Y. P.; Colt, J.; Carver, R.; Burgess, L.; Greenawalt, K. *Proc. Am. Fertil. Soc.* **1994**, *50*, 121.
9. Snibson, G. R.; Greaves, J. L.; Soper, N. D.; Tiffany, J. M.; Wilson, C. G.; Bron, A. J. *Cornea* **1992**, *11*, 288.
10. Dunn, R. L.; Tipton, A. J.; Yewey, G. L.; Reinhart, P. C.; Menardi, E. M.; Rogers, J. A.; Southard, G. L. *Abs. Pap. ACS* **1990**, *200*, 155.
11. Dunn, R. L.; Yewey, G. L.; Duysen, E. D.; Polson, A. M.; Southard, G. L. *Abs. Pap. ACS* **1994**, *208*, 247.
12. Tipton, A. J.; Yewey, G. L.; Dunn, R. L. *Abs. Pap. ACS* **1992**, *203*, 57.
13. Sawhney, A. S.; Pathak, C. P.; Hubbell, J. A. *Macromolecules* **1993**, *26*, 581.
14. Hill-West, J. L.; Dunn, R. C.; Hubbell, J. A. *J. Surg. Res.* **1995**, *59*, 759.
15. West, J. L.; Hubbell, J. A. *React. Polym.* **1995**, *25*, 139.
16. Hill-West, J. L.; Chowdhury, S. M.; Slepian, M. J.; Hubbell, J. A. *Proc. Natl. Acad. Sci. USA* **1994**, *91*, 5967.
17. Slepian, M. J.; Campbell, P. K.; Berrigan, K.; Roth, L.; Massia, S. P.; Weselcouch, E.; Ron, E.; Mathiowitz, E.; Jacob, J.; Chickering, D.; Philbrook, M. *Circulation* **1994**, *90*, 96.
18. Konishi, R.; Firestone, L.; Wagner, W.; Federspiel, W. J.; Konishsi, H.; Shimizu, R.; Hattler, B. *ASAIO J.* **1996**, *42*, 72.
19. Matsuda, T.; Moghaddam, M. J.; Miwa, H.; Sakurai, K.; Iida, F. *ASAIO J.* **1992**, *38*, 154.
20. Barrows, T. H.; Truong, M. T.; Lewis, T. W.;Grussing, D. M.; Kato, K. H.; Gysbers, J. E.; Lamprecht, E. G. *Trans. World Cong. Biomat.* **1996**, *5*, 8.
21. Otani, Y.; Tabata, Y.; Ikada, Y. *J. Biomed. Mater. Res.* **1996**, *31*, 157.
22. Kobayashi, H.; Hyon, S. H.; Ikada, Y. *J. Biomed. Mater. Res.* **1991**, *25*, 1481.

Chapter 8

Transdermal Delivery of Macromolecules: Recent Advances by Modification of Skin's Barrier Properties

Mark R. Prausnitz

School of Chemical Engineering, Georgia Institute of Technology, Atlanta, GA 30332–0100

Great advances in transdermal delivery of macromolecules have been made within the last few years using ultradeformable liposomes, electroporation, and low-frequency ultrasound, each of which has been shown to deliver macromolecules at clinically-useful rates. Transdermal drug delivery is a potentially useful method by which macromolecules, such as proteins, could be administered for local or systemic therapy. Until recently, transdermal delivery was not a realistic option, since skin's great barrier properties had prevented transport of macromolecules across human skin at therapeutically-relevant rates. In this chapter, chemical, electrical, and ultrasonic delivery methods are reviewed. Mechanistic perspective, a summary of key experimental findings, and an assessment of potential for impact on medicine are provided for each enhancement technique.

Biotechnology has produced a generation of novel macromolecular compounds with great therapeutic promise. While a number of challenges sometimes slow progress of these new drugs to clinical application, difficulties in meeting their special drug delivery requirements can be a significant impediment. This is because biologically-active macromolecules, such as proteins, generally have low oral bioavailability, making oral administration difficult, and often have short biological half-lives, making parenteral delivery impractical outside a hospital setting (*1-3*). Delivery of drugs across the skin addresses these problems by offering a number of potential advantages compared to conventional methods, such as pills and injections: (1) no degradation due to stomach, intestine, or first pass of the liver, (2) likely improved patient compliance because of a user-friendly method, and (3) potential for steady or time-varying controlled delivery (*4-9*). However, delivery of therapeutic quantities of macromolecules across human skin is extremely difficult. This chapter describes the current status of transdermal drug delivery, focusing on recent advances involving the modification of skin's barrier properties, which indicate that transdermal delivery of macromolecules (> 1 kDa) may now be possible.

Transdermal drug delivery. The advantages of delivery across skin have led to the clinical success of a number of transdermal products, indicated by annual sales in excess of one billion dollars. Transdermal drugs approved by the United States Food and Drug Administration (FDA) include clonidine, estradiol, fentanyl, lidocaine, nicotine, nitroglycerin, scopolamine, and testosterone (*10*).

Applications of transdermal drug delivery are limited largely by skin's great barrier properties, which prevent transdermal diffusion of most compounds at therapeutic rates (*4-9*). Drugs which have been successfully delivered (i.e., the FDA-approved drugs listed above) each share three common traits: effectiveness at relatively low doses, molecular mass less than 400 Da, and lipid solubility. While proteins and other macromolecular drugs are often effective at low doses, they generally are much larger than 400 Da and have very poor lipid solubility, which explains their extremely slow percutaneous absorption.

Pathways for transport across the skin's stratum corneum. The outer 10 - 15 μm of human skin is the stratum corneum (*11*), a dead layer of tissue which provides the primary and extremely effective barrier to transdermal transport (Figure 1) (*12-14*). Below is the viable epidermis, which consists of living cells, but is devoid of nerves and blood vessels. Deeper still is the dermis, which also contains living cells, in addition to blood vessels and nerves. While most drugs traverse the stratum corneum very slowly, they diffuse with great ease through deeper tissues to the capillary bed in the dermis (*4-9*).

The stratum corneum's barrier properties are generally attributed to multilamellar lipid bilayers which fill the extracellular spaces (*15, 16*). The bulk of stratum corneum is composed of flattened cells called keratinocytes, which are filled with cross-linked keratin. Their relatively permeable cell interiors are not normally accessible for transport, since they are surrounded by the relatively impermeable intercellular lipids. Unlike the phospholipid bilayers of cell membranes, these intercellular bilayers contain very few phospholipids, being composed primarily of ceramides, cholesterol, and fatty acids (*17*).

There are three transport pathways across the skin which molecules are likely to follow (Figure 2). One involves transport directly across the bulk of stratum corneum, where a molecule must sequentially cross keratinocytes and intercellular lipid bilayers. Normally, this route is not available to most molecules, because it involves crossing on the order of 100 intercellular bilayers, which is energetically unfavorable (*18*), and is therefore extremely slow. Another pathway reduces the number of bilayer crossings by following a tortuous path exclusively within the intercellular lipids, where drugs travel predominantly along the multilamellar bilayers, rather than across them. This route is probably taken by small drugs which diffuse across the skin (*19-22*). The third pathway, often termed the "shunt" route, avoids the intercellular lipid bilayers altogether by following a path within sweat ducts and hair follicles. Although the shunts make up only a small fraction of the skin (~ 0.1 % (*12*)), this route is important for transport of charged compounds, especially when electrophoretically driven by an imposed electric field (see below) (*23-27*).

Because stratum corneum lipids limit transport of most compounds, efforts to increase transdermal delivery have often focused on altering lipid bilayer structure to increase permeability. Modification of skin's barrier properties in this way has been achieved by chemical and physical approaches. Recent advances in this field, most of which have been published since 1995, suggest that the tools needed for transdermal delivery of macromolecules are now available.

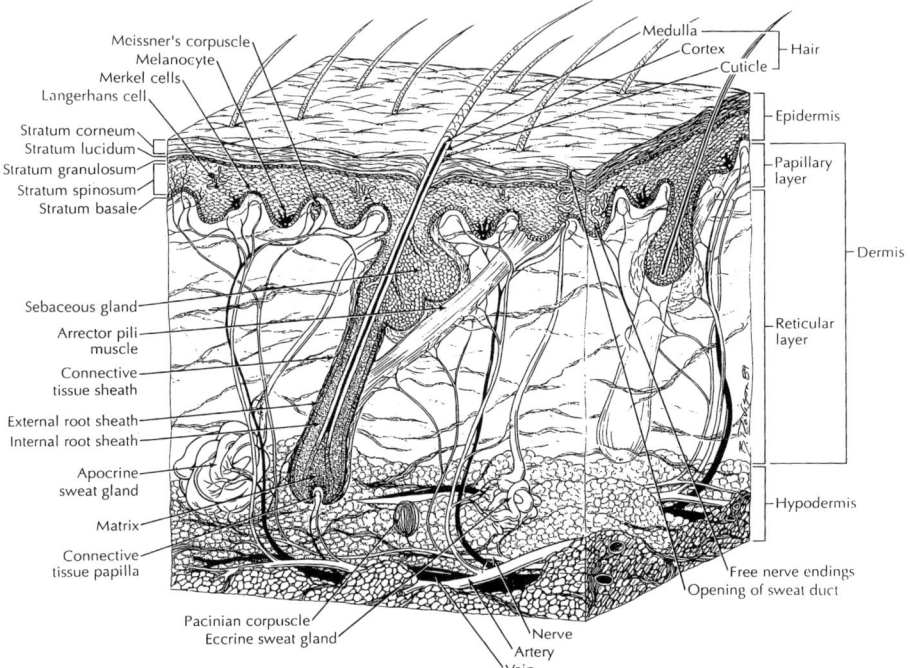

Figure 1. A composite representation of the anatomical structures found in mammalian skin. The outermost layer, stratum corneum, provides the primary resistance to transdermal transport of most compounds. Because the epidermis is avascular, drugs must reach the capillaries (or lymphatic vessels) in the dermis for systemic administration. Reproduced with permission from reference (*13*). Copyright 1991 CRC Press, Inc.

Modification of Skin's Barrier Properties.

Chemical enhancers can alter the skin's lipid environment. Because transport across skin by passive diffusion is too slow for most applications, the effects of a broad variety of chemicals on transdermal drug delivery have been investigated. Although extensively studied, their potential for significant impact on macromolecule delivery has not been demonstrated.

Mechanistic perspective. Transdermal transport by the tortuous intercellular route followed by most drugs can be viewed as a two-step process: (1) drug must first partition from an external donor solution into the skin's lipids and then (2) diffuse across the stratum corneum within the lipid domain. Models based on this approach have successfully described transdermal transport (*19, 22, 28-32*). Chemical enhancers should therefore be effective if they alter the skin's lipid environment in ways which (1) increase drug solubility in skin and/or (2) increase drug diffusivity in skin.

Experimental transdermal permeability values for some hydrophilic compounds are inconsistent with transport via an intercellular lipid route and have led to the hypothesis that there are additional hydrophilic pathways, sometimes called "aqueous pores" (*33-37*). The physical nature of these pathways remains controversial, but may represent hydrophilic domains within the bulk of stratum corneum. Others suggest that diffusion through hair follicles and sweat ducts may be significant (*33, 38, 39*).

Experimental findings. Chemical approaches to increasing transport have received extensive attention from the transdermal community (*4-9, 40, 41*). Most effective chemical enhancers act by disrupting or fluidizing lipid bilayer structures within the stratum corneum, thereby increasing drug diffusivity within the skin. Examples include dimethyl sulfoxide (DMSO), Azone (1-dodecylazacycloheptan-2-one), unsaturated fatty acids (e.g., oleic acid, linoleic acid), and surfactants (e.g. sodium dodecyl sulfate). Chemical enhancers have been shown to increase transdermal transport of small compounds by as much as orders of magnitude, but also frequency cause signifcant skin irritation and may affect drug stability (*4-9, 40, 41*). However, studies addressing chemical enhancement of macromolecules report only modest or no enhancement under clinically-relevant conditions (*42-50*).

Potential for impact. Despite extensive research, chemical enhancers have so far had little practical impact on transdermal delivery beyond preclinical studies. While ethanol is used in FDA-approved formulations (*10*), other enhancers with much greater effects on skin permeability have not yet found clinical acceptance due largely to safety concerns and the costly FDA approval process. Moreover, although transdermal delivery of small molecules is significantly increased by a number of different chemical additives, delivery of proteins and other macromolecules generally is not.

Liposomes facilitate transdermal transport by a poorly understood mechanism. Encapsulation of drugs within liposomes has been studied for many drug delivery applications, including transdermal delivery. Liposomes are spherical lipid bilayer membranes which surround an aqueous interior. In addition to being found in hundreds of cosmetic formulations, liposomes are currently used to enhance transdermal transport of low molecular weight drugs in some pharmaceutical products (*51-54*). Recent laboratory studies indicate that liposomes may also play a useful role in transdermal delivery of macromolecules.

Mechanistic perspective. It is again useful to consider transdermal transport as a two-step process, involving partitioning and diffusion. The ability of liposomes to facilitate drug partitioning into skin is generally accepted (*51-54*). Liposomes can be

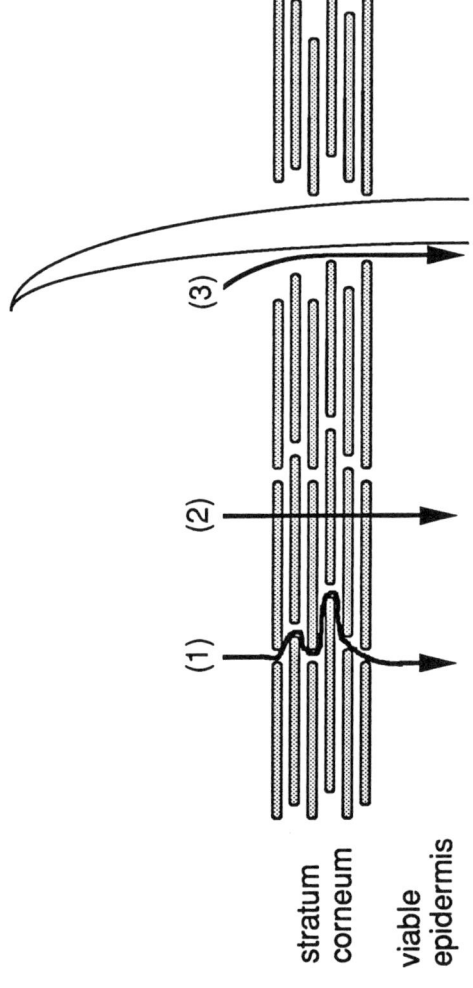

Figure 2. Three primary routes across the stratum corneum are available for transport. (1) A tortuous intercellular route solely within the stratum corneum lipids is usually followed by small compounds diffusing through normal or chemically-modified skin. (2) Transcellular pathways, which cross both the cells and intercellular lipids of stratum corneum, may only be accessible following skin permeabilization by electroporation or ultrasonic cavitation. (3) "Shunt" pathways through the hair follicles and/or sweat ducts may be important during iontophoretic and liposomal transport.

used with lipophilic drugs, which localize within the liposome's lipid bilayer shell, and with hydrophilic drugs, which localize within the aqueous interior. In either case, an interaction of liposomes with the lipids of the stratum corneum could increase drug entry into the skin. Partitioning could be enhanced by liposomes which provide a high local drug concentration at the skin surface. Adsorption or fusion of liposomes onto stratum corneum lipid bilayers could also promote drug partitioning into skin.

The role of liposomes in enhancing the second step in the transport process — diffusion across the stratum corneum — remains controversial. Most researchers agree that conventional liposomes do not serve as drug carriers which cross the bulk of stratum corneum as intact vesicles (54-56). Enhanced drug diffusion within stratum corneum could result from liposomal lipids becoming incorporated into stratum corneum bilayers, thereby acting as chemical enhancers which fluidize or otherwise change lipid properties to facilitate transport (57).

Under special circumstances, some suggest that liposomes cross the stratum corneum as intact vesicles. The hair follicles may provide a shunt route through which liposomes could cross the stratum corneum and deposit drug deep within the skin, primarily within or near hair follicles (58, 59). Moreover, liposome formulations designed to make vesicle shape very deformable might penetrate intact skin more readily (60, 61).

Experimental findings. Liposomes have been shown to increase topical and transdermal delivery of a variety of low molecular weight compounds (51-54). Moreover, it has been shown that liposomes can increase drug localization within the skin while decreasing systemic distribution (62-64). This has made the use of liposomes a popular enhancement technique for local drug delivery in dermatological applications.

Delivery of macromolecules can also be enhanced by liposomes, where transport is usually localized within hair follicles. Increased macromolecule penetration into skin has been shown following topical administration with liposomes for cyclosporin (1.2 kDa) (62, 65), a DNA repair enzyme (16 kDa) (66), γ-interferon (16 - 25 kDa, in monomeric form) (67), α-interferon (18 - 20 kDa) (65), melanin (68), superoxide dismutase (33 kDa) (69), a monoclonal antibody (~150 kDa) (70), and DNA (1 kb) (71). These studies generally used animal skin in vivo, in vitro, or from histoculture. Some studies have directly shown localization of these compounds within follicles, while others have inferred it. It has not been clearly shown that intact liposomes penetrate deep into follicles without breaking up. Moreover, some studies indicate that molecules need not be encapsulated within liposomes for increased follicular penetration, but can be co-administered in solution (72). This suggests that intact liposomes may not carry drug across the skin, but enhance transport by a different mechanism.

To facilitate liposome penetration into skin, ultradeformable liposomes (termed "transfersomes") have been developed through the addition of bile salts to liposome bilayers (60, 61). Enhanced transport of a number of small drugs has been shown using this approach, including the clinical delivery of lidocaine to increase local anesthesia (73). Studies which demonstrate systemic delivery of macromolecules across the skin have used insulin (5.8 kDa, in monomeric form) (74), bovine serum albumin (69 kDa) (75), and gap junction protein (> 178 kDa) (75) (Figure 3). These liposomes must be applied non-occlusively so that the formulation dries, thereby potentially enhancing an osmotic driving force for transport (60). It has been proposed that ultradeformable liposomes cross the skin as intact vesicles, following a non-follicular pathway and being taken up by the lymphatic system before entering systemic circulation (61).

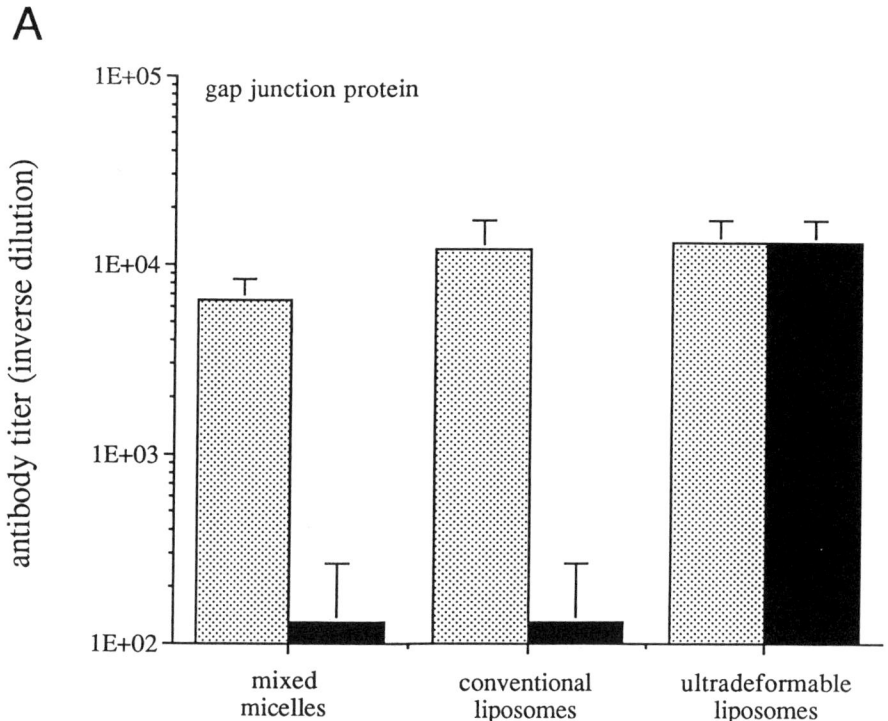

Figure 3. Immune response to (A) gap junction protein or (B) bovine serum albumin given to mice either by injection (▒) or topical application (■). Immunogenic protein was administered with mixed micelles (soybean phosphatidylcholine (SPC) and bile salt at 1:1 mole ratio) or encapsulated within liposomes, either "conventional" (SPC only) or ultradeformable (SPC and bile salt at 9:2 mole ratio). Antibody titers were determined by the serum-dependent, complement-mediated lysis of antigen-sensitized liposomes. Topically applied protein elicited an immune response equal to that caused by injection only when administered with ultradeformable liposomes. Standard deviation bars are shown. Reproduced with permission from reference (75).
Copyright 1995 VCH Verlagsgesellschaft mbH.

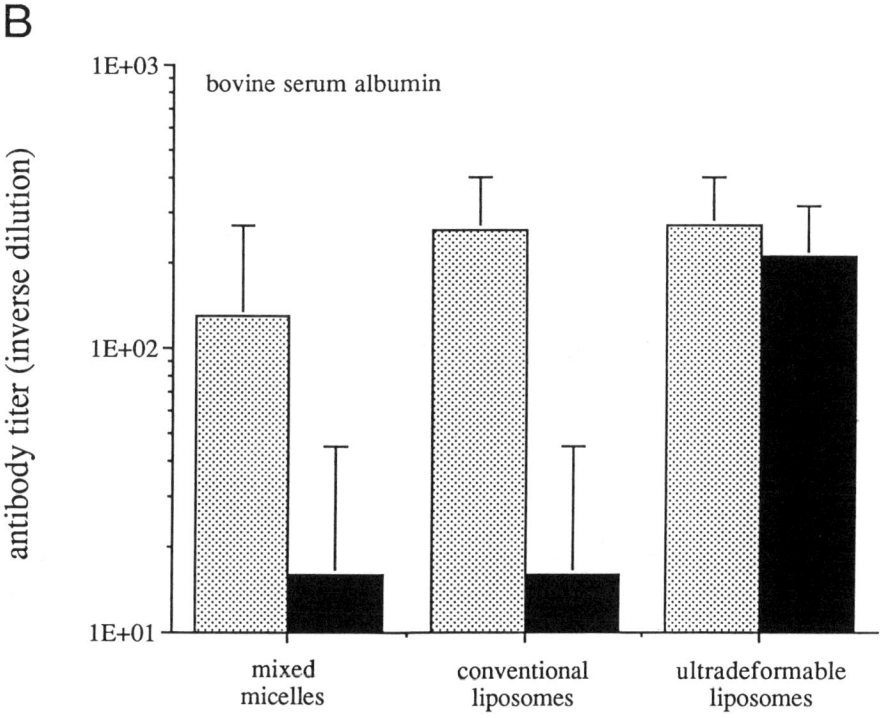

Figure 3B. *Continued*

Potential for impact. Although only a limited number of studies address liposomal delivery of macromolecules, they demonstrate penetration of large compounds into or across skin. Topical delivery of macromolecules to hair follicles has a number of potential clinical applications. Reports on systemic macromolecule delivery using ultradeformable liposomes offer still more possibilities.

Iontophoresis provides an electrical driving force for transdermal transport.
As an alternative to enhancing transport by modifying the chemical environment in skin, the possibility of driving drugs across skin by the application of an electric field has received considerable attention, especially since the mid 1980s. Extensive in vitro, in vivo, and clinical study, coupled with a few commercial products, suggest that iontophoresis is a viable means of enhancement. Nevertheless, therapeutic delivery of macromolecules across human skin has been difficult to achieve.

Mechanistic perspective. Application of an electric field across the skin can enhance transdermal transport of both charged and uncharged drugs by electrophoresis and/or electroosmosis. A charged compound in an electric field moves by electrophoresis at a rate determined by the product of the electric field strength and the compound's mobility (a function of molecular size and charge) in the surrounding medium (i.e., the skin) (76). Electrophoretic enhancement is often possible, since many drugs have a net charge, including most macromolecules of therapeutic interest.

Because the skin carries a net fixed negative charge, transdermal transport of positively-charged ions is favored. As a result, during iontophoresis there is a net flux of ions from the anode to the cathode, which provides a convective driving force for transport across the skin, termed electroosmosis (77). With the proper electric field orientation, this effect can be used to enhance transport of uncharged compounds. Moreover, positively-charged drugs delivered across the skin by electrophoresis will be further enhanced by electroosmosis. However, electroosmosis will oppose electrophoretic transport of negatively-charged drugs. Theoretical models have been developed which describe transdermal electrophoresis and electroosmosis (22, 77-79). While some studies suggest that during iontophoresis ions cross the stratum corneum via the tortuous intercellular routes followed during passive diffusion (80, 81), others identify appendageal shunts as the primary pathways (23-27).

Electrical studies show that exposure of skin to transdermal voltages of approximately one volt or more reduces skin resistance. At typical iontophoretic voltages (< 10 V), human skin resistivity drops one to two orders of magnitude (from 100 kΩ-cm^2) over a timescale of seconds to tens of minutes (24, 27, 82-88). Lowered skin resistance and increased skin permeability can persist after the electric field is removed, demonstrating either partial or full reversibility over a timescale of minutes to hours. Mechanistically, this has been explained by voltage-dependent rearrangements in skin microstructure (79, 87) and by an electroosmotic mechanism (85). These electrical measurements suggest that electric fields are capable of not only driving molecules across skin, but directly changing skin barrier properties. The possibility of utilizing electric fields in this way has received some attention (24, 77, 87), with recent efforts directed outside the context of traditional iontophoresis, as described below (see Electroporation).

Experimental findings. Transdermal iontophoresis of small compounds has been the subject of extensive in vitro and in vivo studies, some of which have led to clinical success (89-92). Today, commercial products exist for iontophoresis of: pilocarpine to induce sweating as part of a cystic fibrosis screening test (e.g., CF Indicator, Medtronic, Inc., Minneapolis, MN) (93), tap water as a treatment for hyperhidrosis (e.g., Drionic, General Medical Co., Los Angeles, CA) (94), and lidocaine and other therapeutic agents from an "all-purpose" device (e.g., Iontocaine and Phoresor, IOMED, Inc., Salt Lake City, UT) (*95*). Iontophoresis is generally well tolerated, although mild skin irritation, erythema, and non-painful sensation are sometimes reported.

Iontophoresis of macromolecules has been more difficult than electrically-assisted delivery of small compounds (*6, 8, 96*). Those reporting success have generally not used human skin, but have demonstrated transport across animal skin, which is often more permeable than the human integument. Macromolecules delivered across animal skin include: arginine-vasopressin (1.1 kDa) (*97*), leuprolide (1.2 kDa) (*98, 99*), calcitonin (3.5 kDa) (*100*), growth hormone releasing factor (3.9 kDa) (*101*), carboxy-inulin (5.2 kDa) (*84*), insulin (*102-105*), and bovine serum albumin (*84*) (Figure 4). Notable exceptions include the clinical delivery of leuprolide to human subjects (*106, 107*) and delivery of cytochrome c (12 kDa) across human skin in vitro (*182, 183*). Delivery of leuprolide, a cholecystokinin-8 analogue (1.2 kDa), and insulin were also achieved across human skin in vitro, where detectable fluxes were measured only when iontophoresis was preceded by a two hour exposure to absolute ethanol (*108, 109*), a process unlikely to find clinical acceptance.

Potential for Impact. Overall, the iontophoresis literature shows that despite success with animal skin, clinically-relevant iontophoretic protocols have been capable of transporting macromolecules across human skin in very few cases. Perhaps an iontophoretic driving force coupled with a means of reversibly altering the skin's barrier properties (e.g., chemical, electrical, or ultrasonic) will be more broadly useful.

Electroporation creates new transdermal pathways by disrupting lipid bilayer structure. Short, high-voltage pulses which cause electroporation are known to transport large numbers of macromolecules across cell membranes without killing cells in vitro and in vivo. Recently, electroporation of the stratum corneum's lipid bilayers was shown to occur and to increase rates of transdermal transport by orders of magnitude (*110, 111*). Subsequent studies indicate that electroporation can also enhance transdermal macromolecule delivery to clinically-relevant levels.

Mechanistic perspective. Electroporation (also called electropermeabilization) is believed to involve the creation of transient aqueous pathways in lipid bilayers by the application of a short (μs to ms) electric field pulse (*112-114*). Permeability and electrical conductance of lipid bilayers are rapidly increased by many orders of magnitude, where membrane changes can be reversible or irreversible, depending mainly on pulse magnitude and duration. This phenomenon is known to occur when the transmembrane voltage reaches approximately 1 V for electric field pulses typically of 10 μs to 100 ms duration when applied to bilayers in either living cells or metabolically-inactive systems (e.g., liposomes). During electroporation the following sequence of events is believed to take place: (1) new aqueous pathways ("pores") are created on a timescale of microseconds or less, (2) molecules are moved through these pathways by diffusion and local electrophoresis and/or electroosmosis, and (3) after the pulse, pores close over characteristic times ranging from milliseconds to hours.

Because the rate-limiting barrier to transdermal transport is the lipid bilayers of the stratum corneum, electroporation of these bilayers could significantly increase drug delivery across skin. A simple theoretical estimate indicates that millisecond electric field pulses of approximately 100 V could electroporate the approximately 100 multilamellar bilayers crossed in a path directly across the stratum corneum (*111, 115, 116*). Voltages typically applied during iontophoresis (< 10 V) are considerably lower, but might be sufficient to electroporate a few bilayers, perhaps affecting the lining of appendages (*79, 87*). Experiments investigating skin electroporation usually apply a

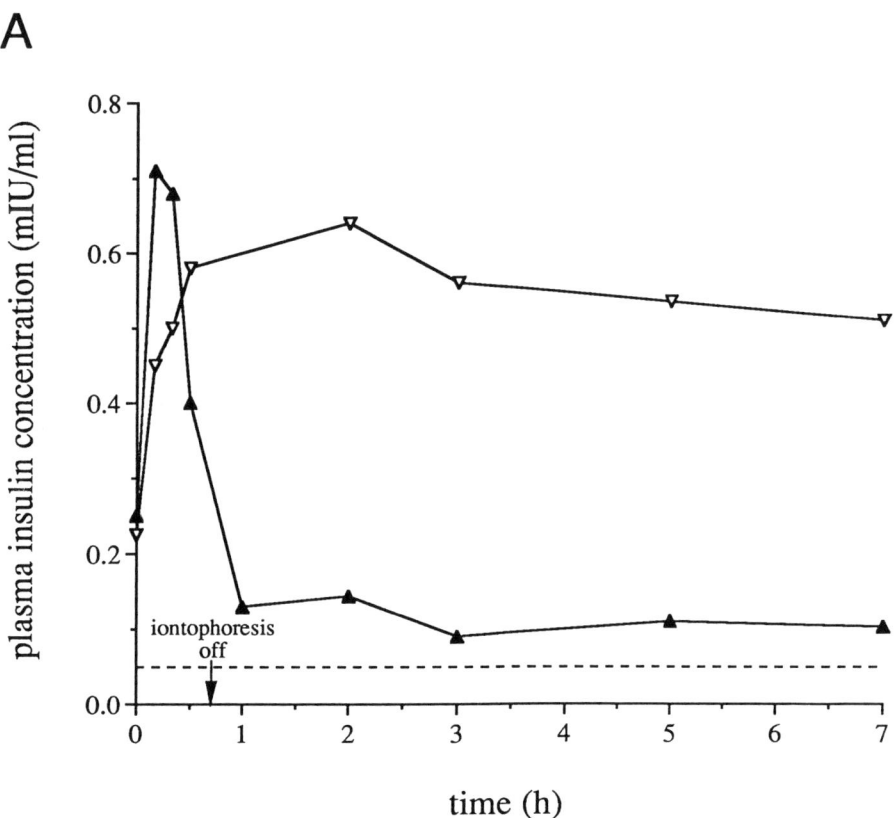

Figure 4. (A) Plasma insulin and (B) blood glucose concentrations following insulin administration to rabbits. Diabetic rabbits were given insulin either by subcutaneous injection (▽) or transdermal administration using iontophoresis (▲). As controls, additional diabetic (O) and normal (□) rabbits received no treatment. Iontophoresis was applied at 1 mA for 40 min using a pulsed waveform. Insulin concentration was determined by radioimmunoassay. Standard deviation bars are shown. Reproduced with permission from reference (96).
Copyright 1990 Elsevier Science - NL., The Netherlands.

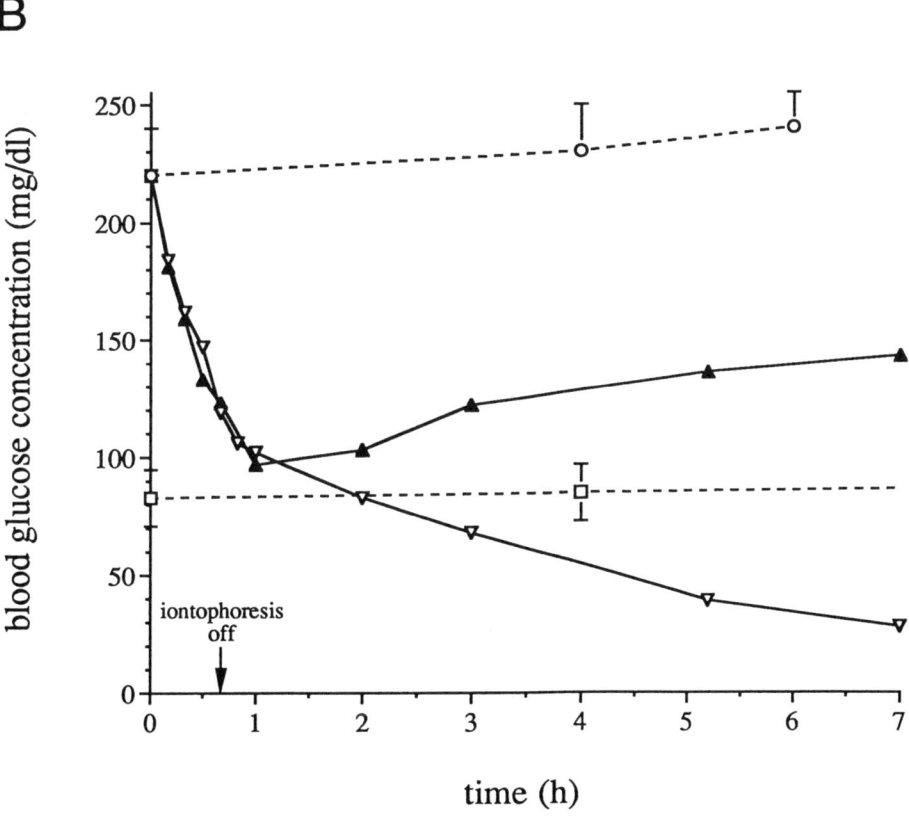

Figure 4B. *Continued*

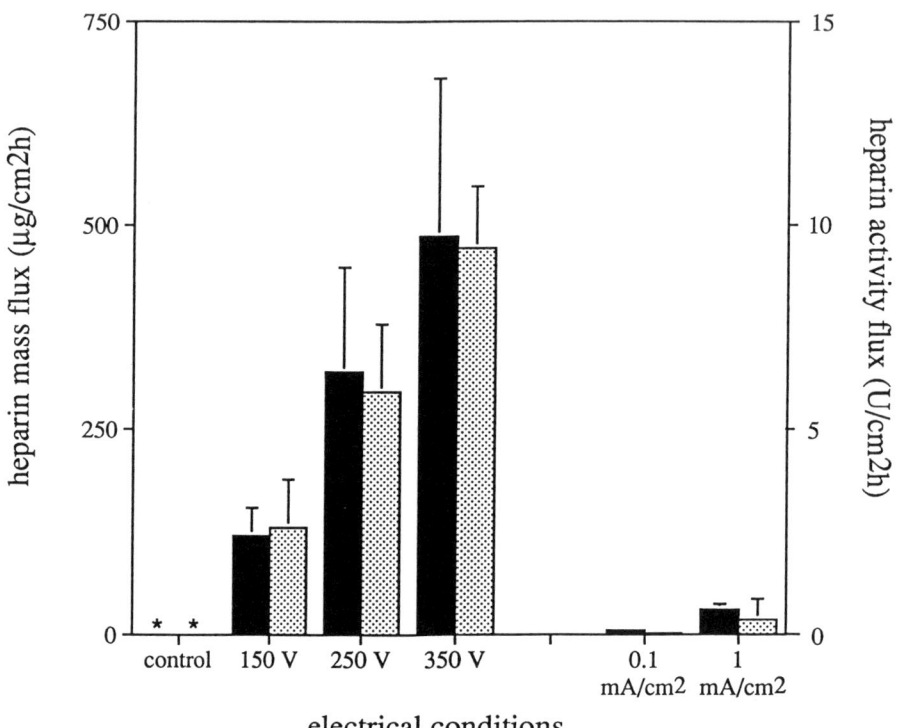

Figure 5. Transdermal heparin fluxes caused by electroporation or iontophoresis of human skin in vitro. Heparin mass flux (■) was determined by scintillation counting measurements of radioactively-labeled heparin. Biological activity flux (▩) was determined by a blood clotting time assay. Heparin fluxes caused by electroporation may be sufficient for clinical applications (126). Electrical exposures were each 1 h of either continuous iontophoresis (0.1 or 1 mA/cm^2) or intermittent electroporation pulses (150, 250, or 350 V) each lasting 1.9 ms and applied at a rate of 12 pulses per minute. Standard deviation bars are shown. Asterisk indicates a flux below the detection limit (of order 1 μg/cm^{2}h for radioactivity measurements and 0.1 U/cm^{2}h for biological activity measurements). Reproduced with permission from reference (126). Copyright 1995 Nature Publishing Company.

series of pulses at rates of one pulse every five seconds to one pulse every five minutes, where each pulse causes a transdermal voltage of 30 to 300 V and lasts for 1 to 300 ms. Theoretical studies suggest that the described experimental findings could be accounted for by electroporation of stratum corneum lipids (*115, 116*).

Experimental findings. Electrical studies have shown that short, high-voltage pulses can have dramatic and reversible effects on skin electrical properties. During a pulse, skin resistance drops as much as three orders of magnitude within microseconds (*117, 118*). Skin resistance then generally recovers by a factor of ten within milliseconds, and exhibits either complete or partial reversibility within minutes. Skin capacitance has been observed to increase by up to an order of magnitude and later reverse to pre-pulse values (*117, 119*). Increased capacitance may indicate changes in skin lipids, since skin's capacitance is generally attributed to stratum corneum lipid bilayers (*80, 86*). In contrast, these electrical effects are not observed during low-voltage iontophoresis (*82, 85-87, 120, 121*).

High-voltage pulses also change skin transport properties such that up to 10,000-fold increases in transdermal delivery occur for compounds ranging in size from small ions to microspheres (*111, 118, 122-136*). Steady-state transport can be achieved in a matter of minutes (*130, 137*). Complete or partial reversibility is generally observed within an hour (*111, 130*). Microscopic imaging suggests that transport occurs through the bulk of stratum corneum via transcellular and intercellular pathways estimated to occupy up to 0.1% of skin area (*138, 139*). This observed transcellular transport contrasts with the tortuous intercellular pathways of passive diffusion and the shunt pathways of iontophoresis and liposomes. Limited work performed on hairless rats indicates that large transdermal fluxes are also achieved in vivo, where no effects beyond transient erythema and edema were observed (*111, 123, 140*). Additional safety studeis are required.

The effects of high-voltage pulses have been attributed to short-lived changes in skin structure (e.g., electroporation of stratum corneum lipid bilayers) followed by electrophoretically-driven transport across the skin. A number of studies indicate that electrophoresis alone cannot explain the large flux increases observed during high-voltage pulsing, supporting the hypothesis that skin structure is transiently disrupted (*111, 134, 136, 141*). Other studies more specifically suggest that high-voltage pulses can create enlarged transport pathways, the size of which is controlled by a voltage-dependent mechanism (*118, 126*).

Skin electroporation can increase delivery of macromolecules to therapeutically-useful rates across human skin. Transdermal transport of heparin (5 - 30 kDa) was increased by electroporation in vitro to rates sufficient for clinical anticoagulation therapy (*126*) (Figure 5). Other studies have demonstrated electroporation-enhanced delivery of arginine-vasopressin (*142*), luteinizing hormone releasing hormone (1.2 kDa) (*122, 140, 142*), neurotensin (1.7 kDa) (*142*), and oligonucleotides (4.8 and 7 kDa) (*129*). Increased penetration into skin has also been shown for latex microspheres of up to micron dimensions (*125, 139*). These studies almost all employed human skin in vitro.

Potential for impact. Electroporation's ability to both create new transport pathways and drive molecules through them has proved capable of delivering macromolecules across skin, especially for highly-charged compounds which are effectively moved by electrophoresis (e.g., heparin, oligonucleotides). This has the potential to lead to a variety of clinical applications. Electroporation-mediated delivery of macromolecules which carry a weak net charge, such as proteins, has not yet received attention. Issues of safety and drug stability also need further study.

Ultrasound creates new transdermal pathways by cavitation. Ultrasound is used extensively in clinical practice for applications ranging from diagnostic imaging to therapeutic heating to lithotripsy procedures. Transdermal delivery enhanced by ultrasound (sometimes called sonophoresis or phonophoresis) has received sporadic attention for half a century, but has recently sparked renewed interest by studies which demonstrate delivery of macromolecules at therapeutically-relevant rates (*143*).

Mechanistic perspective. Ultrasound is a pressure wave having a frequency too high to be heard by the human ear (> 16 kHz) (*144, 145*). When introduced into the body, ultrasound echoes off internal structures, thereby allowing diagnostic imaging. Ultrasound conditions used by diagnostic instruments are typically very high frequency (» 1 MHz) and low intensity (« 1 W/cm^2), selected in part to prevent damaging imaged tissue (*144, 146-150*). Ultrasound under these conditions should have no effect on skin properties.

Ultrasound applied at "therapeutic" conditions heats tissue. High frequencies (~ 1 MHz) and moderate intensities (~ 1 W/cm^2) are typically employed in physical therapy and cancer chemotherapy using ultrasonic hyperthermia (*144-150*). These conditions are favorable because they provide sufficient energy to heat tissue, even deep within the body, without causing other effects associated with ultrasound at greater intensity or lower frequency. Most studies using ultrasound to enhance transdermal drug delivery have used therapeutic conditions (*151-156*), in part because those intensities and frequencies are already FDA-approved for clinical use. Ultrasonic heating of skin could increase transdermal transport by fluidizing stratum corneum lipids and/or increasing convective flow.

Ultrasound can also cause non-thermal effects such as cavitation. If applied at lower frequencies (« 1 MHz) or greater intensities (» 1 W/cm^2) than used in therapeutic applications, ultrasound can cause extensive generation of gas bubbles, called cavitation (*144-150, 157, 158*). Stable cavitation creates bubbles which oscillate in size at the frequency of the applied ultrasound. Transient cavitation bubbles are short lived, imploding violently upon their collapse. Both forms of cavitation can have severe effects on biological tissue, as demonstrated by the shattering of kidney stones during ultrasonic lithotripsy procedures (*150, 159, 160*) and ultrasonic cell disruption techniques commonly employed in research laboratories (*150, 161*). If applied to the skin, significant changes in skin structure and permeability could result.

Experimental findings. Ultrasound has been used since the 1950's to enhance transport of small drugs into and across skin for local delivery. Early clinical studies showed increased absorption of hydrocortisone when accompanied by ultrasound at therapeutic intensity and frequency (*162, 163*). Other clinical studies have described local delivery of anesthetics, non-steroidal anti-inflammatories, antibiotics, and antivirals (*151-156*). In addition to local delivery, systemic transdermal administration of small drugs has also been enhanced by ultrasound. While some work has been performed clinically (*164*), most compounds have been examined through in vitro and in vivo laboratory investigation (*165-170*). These studies, in addition to others performed outside the context of drug delivery, suggest that application of therapeutic ultrasound to the skin is safe (*146, 149, 150*).

The mechanism(s) of ultrasonic enhancement remains controversial. While most conclude that thermal contributions are small, evidence for increased convection (e.g., acoustic streaming or mixing) (*165, 171*) and for cavitation-mediated effects (*165, 170, 172*) offer compelling explanations for increased transport. Although many studies have shown that therapeutic ultrasound can significantly enhance transdermal transport (*151-156*), some studies report that ultrasound has no effect (*173-175*). A recent analysis may reconcile these findings by identifying that studies which report no enhancement used compounds which are small (< 250 Da), while those which observed enhanced transport used larger compounds (*176*). Because small molecules diffuse easily through stratum corneum lipids, their transport may not be significantly increased by increases in lipid fluidity caused by ultrasound. Concerning the route of transport during therapeutic ultrasound, studies suggest that compounds follow an intercellular path (*168*).

Deviating from conventional therapeutic ultrasound, a few studies have employed lower frequencies (e.g., 20 - 100 kHz), which cause increased cavitation (*145, 177*). Enhanced delivery of lidocaine (*178*) and insulin (*172, 179*) has been shown in animal models. Recently, transdermal delivery of insulin, γ-interferon, and erythropoeitin (48 kDa) was demonstrated using human skin in vitro, supported by data obtained with hairless rats in vivo (*143*) (Figure 6). Delivery of these large macromolecules was at rates sufficient for clinical applications. Preliminary histological examination showed no adverse effects (*143, 180*). In contrast to therapeutic ultrasound, the effects of low-frequency ultrasound are believed make transcellular pathways accessible by a mechanism involving cavitation (*180*).

Potential for impact. Although useful for local and systemic delivery of small compounds, therapeutic ultrasound has not significantly increased transdermal transport of macromolecules. When applied at lower frequencies, ultrasound has been shown to deliver large macromolecules across skin at therapeutically-relevant rates. Although only limited work has been done so far, the dramatic effects of low-frequency ultrasound suggest that it is a promising new approach which warrants close attention. Safety and drug stability concerns will require further study.

Discussion.

Methods of delivering macromolecules across skin which have been successful each provide a driving force for transport as well as modify skin's barrier properties. Methods which do not alter skin's properties, such as passive diffusion and iontophoretic enhancement by electrophoresis and/or electroosmosis, have weaker ability to transport large compounds. In contrast, barrier modification by means of electroporation or ultrasonic cavitation increases macromolecule delivery to levels of clinical interest. Disruption of the skin with chemicals can sometimes increase macromolecule delivery, but generally only under conditions which raise safety concerns. Liposomes are an exception, since they deliver macromolecules by a mechanism which is not yet completely understood, but appears not to involve modifying the skin barrier.

Introduction of electrical or ultrasonic energy into skin alters skin properties as a complex function of the energy input. As shown in Table I, iontophoresis, electroporation, therapeutic ultrasound and low-frequency ultrasound all introduce energy into the skin, but have very different effects on the skin barrier. Increases in macromolecule transport do not relate in a simple manner to the energy provided to the skin. For example, therapeutic ultrasound supplies the greatest average power, but has the weakest effect. While the instantaneous power associated with electroporation is 100-times greater than ultrasound, low-frequency ultrasound appears to deliver large macromolecules more effectively. Enhancement also does not relate simply to changes in skin resistance. The effects of low-frequency ultrasound on skin resistance are similar to those caused by iontophoresis, yet the abilities of these two methods to enhance macromolecule transport are very different. Clearly, the form of energy provided, and its microscopic distribution within the skin, are important. Perhaps it is the delivery of short doses of highly localized energy that is most effective. This is achieved during electroporation, where millisecond-pulses of energy are concentrated at

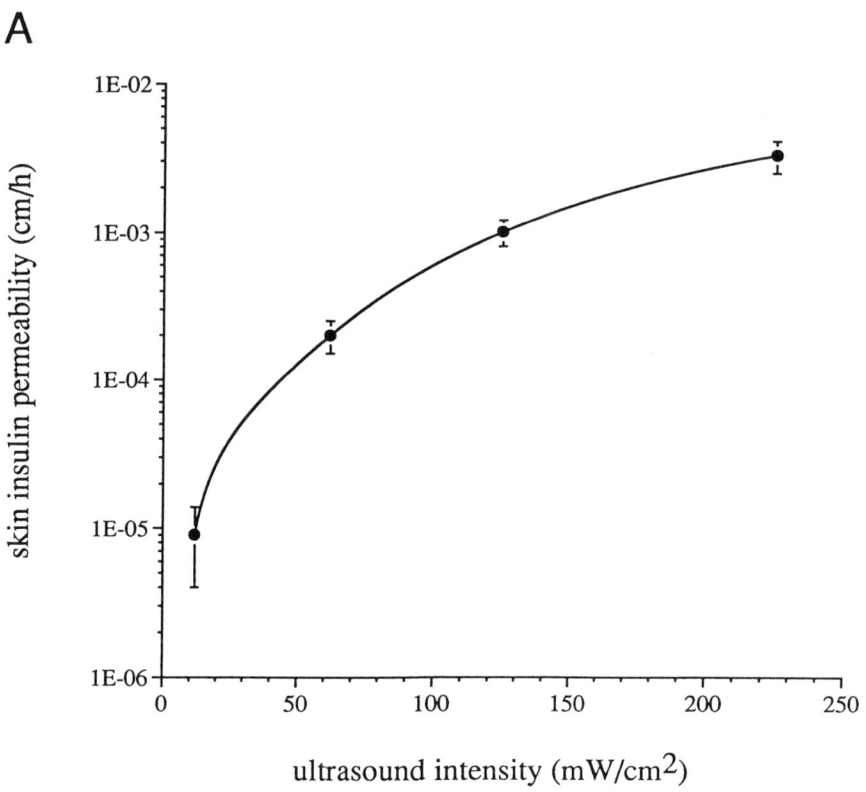

Figure 6. Transdermal delivery of insulin in the presence of ultrasound. (A) The permeability of human skin in vitro to insulin as a function of ultrasound intensity (at 20 kHz and 10% duty cycle). A permeability on the order of 10^{-3} cm/h may be sufficient to deliver insulin at clinically-useful rates (143). (B) Blood glucose concentration in hairless rats. The blood glucose level in diabetic rats was reduced to normal levels by application of ultrasound (225 mW/cm^2, 20 kHz, 10% duty cycle, 30 min) to an insulin solution on the rat's skin (▲). The blood glucose levels of diabetic (O) and normal (□) rats which received neither insulin nor ultrasound remained constant. Standard deviation bars are shown. Reproduced with permission from reference (143).
Copyright 1995 American Association for the Advancement of Science.

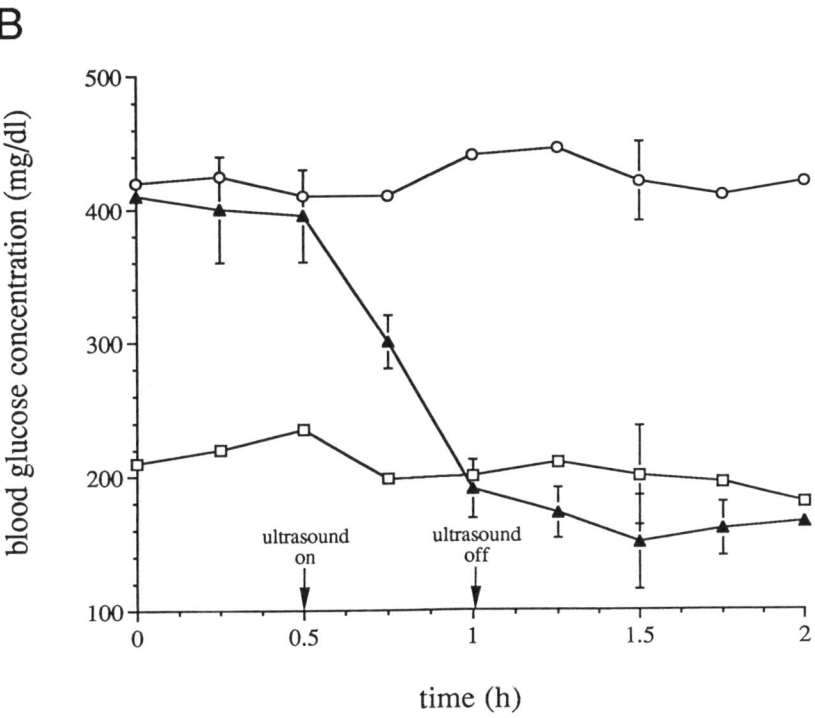

Figure 6B. *Continued*

Table I. Characteristics of ultrasonic and electrical methods of transdermal transport enhancement.[a]

characteristic properties	ultrasound (therapeutic)	iontophoresis	electroporation	ultrasound (low-frequency)
instantaneous power[b] (W/cm^2)	1	0.001	100	1
average power[b] (W/cm^2)	1	0.001	0.001 - 0.1	0.1
skin resistance[c] (kΩ-cm^2)	100	1 - 10	0.1	1 - 10
timescale of resistance change[c]	1 h	1 - 10 min	1 μs	1 h
macromolecule transport[d]	not likely	possible	possible	likely

[a] Numerical values provided in this table are based on data from references (88, 118, 180).
[b] Iontophoresis is typically applied continuously, while pulsed fields are often used in ultrasound (e.g., 10% duty cycle) and electroporation (0.001 - 0.1% duty cycle). A typical iontophoresis exposure could apply up to a few volts and ≤ 0.5 mA/cm^2 across the skin. Electroporation typically applies ~100 V and ≤ 1 A/cm^2 instantaneously during pulses.
[c] Exposure to ultrasound and electric fields can lower skin resistance. Representative steady state resistance values and the timescales over which those values are reached are shown for human skin.
[d] The possibility of delivering macromolecules across human skin at clinically-relevant rates is assessed, based on information presently available in the literature.

the lipid bilayers, and during low-frequency ultrasound, where cavitation releases localized bursts of energy within the skin.

Recognizing that successful macromolecule delivery involves two important components — (1) modifying the skin barrier and (2) providing a driving force for transport — combinations of enhancement methods may provide new opportunities. Used alone, electroporation and ultrasound inherently affect both of these components, where electroporation simultaneously disrupts the skin and transports molecules by electrophoresis and/or electroosmosis, and ultrasound disrupts the skin via cavitation and enhances transport by acoustic streaming and other convective effects. An example where two different methods have been combined, iontophoresis has been used to provide a driving force for transport across skin exposed to chemical enhancers which alter the skin barrier (*108, 109*). Similarly, electrophoretic transport by iontophoresis has been preceded by electroporation, used to permeabilize the skin (*140, 141*). Therapeutic ultrasound has also been applied during electroporation to enhance the ability of electroporation to disrupt the skin barrier (*181*).

Conclusion.

Research on transdermal transport has led to a number of successful clinical applications involving delivery of low molecular weight drugs. However, the special delivery requirements of proteins and other therapeutic products of biotechnology present the need to expand the scope of transdermal administration to include delivery of macromolecules. Just a few years ago, this need could not be met, since transdermal delivery of macromolecules across human skin at clinically-relevant rates had not been performed. However, successful macromolecule delivery has now been demonstrated, in which transient modification of skin's barrier properties is an important component. Techniques involving ultradeformable liposomes, electroporation, and low-frequency ultrasound, perhaps supplemented with chemical or iontophoretic enhancement, show real promise as tools for delivering biologically-active macromolecules across skin.

Despite the great importance of developing drug delivery technologies for macromolecules, most transdermal delivery research continues to focus on traditional methods of enhancement useful primarily for small compounds. With few exceptions, the research groups which first reported the effects of ultradeformable liposomes, electroporation, and low-frequency ultrasound remain the only ones who have published on the subject. Contributions by other researchers would confirm their exciting results and provide new perspectives on the rapidly evolving field of transdermal macromolecule delivery.

Acknowledgments.

Thanks to Mark Johnson and Samir Mitragotri for helpful discussions and to Nancy Monteiro-Riviere for providing Figure 1. This work received partial support from the Whitaker Foundation and the Hoechst-Celanese Corporation. This chapter is based in part on a paper in *Critical Reviews in Therapeutic Drug Carrier Systems*.

References.
1   Robinson, J. R.; Lee, V. H., Eds. *Controlled Drug Delivery: Fundamentals and Applications*; Marcel Dekker: New York, 1988.
2   Kost, J., Ed. *Pulsed and Self-Regulated Drug Delivery*; CRC Press: Boca Raton, FL, 1990.

3. Cleland, J. L.; Langer, R., Eds. *Formulation and Delivery of Proteins and Peptides*; American Chemical Society: Washington, DC, 1994.
4. Bronaugh, R. L.; Maibach, H. I., Eds. *Percutaneous Absorption, Mechanisms — Methodology — Drug Delivery*; Marcel Dekker: New York, 1989.
5. Hadgraft, J.; Guy, R. H., Eds. *Transdermal Drug Delivery: Developmental Issues and Research Initiatives*; Marcel Dekker: New York, 1989.
6. Cullander, C.; Guy, R. H. Transdermal delivery of peptides and proteins. *Adv. Drug Deliv. Rev.* **1992**, *8*, 291-329.
7. Hsieh, D. S., Ed. *Drug Permeation Enhancement*; Marcel Dekker: New York, 1994.
8. Amsden, B. G.; Goosen, M. F. A. Transdermal delivery of peptide and protein drugs: an overview. *AIChE J.* **1995**, *41*, 1972-1997.
9. Smith, E. W.; Maibach, H. I., Eds. *Percutaneous Penetration Enhancers*; CRC Press: Boca Raton, FL, 1995.
10. *Physicians' Desk Reference*; Medical Economics Data Production Company: Montvale, NJ, 1996.
11. Holbrook, K. A.; Odland, G. F. Regional differences in the thickness (cell layers) of the human stratum corneum: an ultrastructural analysis. *J. Invest. Derm.* **1974**, *62*, 415-422.
12. Bronaugh, R. L.; Stewart, R. F.; Congdon, E. R. Methods for in vitro percutaneous absorption studies II. Animal models for human skin. *Toxicol. Appl. Pharmacol.* **1982**, *62*, 481-488.
13. Monteiro-Riviere, N. A. Comparative anatomy, physiology, and biochemistry of mammalian skin. In *Dermal and Ocular Toxicology*; Hobson, D. W., Ed.; CRC Press: Boca Raton, FL, 1991; pp. 3-71.
14. Champion, R. H.; Burton, J. L.; Ebling, F. J. G., Eds. *Textbook of Dermatology*; Blackwell Scientific: London, 1992.
15. Bouwstra, J. A.; de Vries, M. A.; Gooris, G. S.; Bras, W.; Brussee, J.; Ponec, M. Thermodynamic and structural aspects of the skin barrier. *J. Control. Release* **1991**, *15*, 209-220.
16. Elias, P. M. Epidermal barrier function: intercellular lamellar lipid structures, origin, composition and metabolism. *J. Control. Release* **1991**, *15*, 199-208.
17. Lampe, M. A.; Burlingame, A. L.; Whitney, J.; Williams, M. L.; Brown, B. E.; Roitman, E.; Elias, P. M. Human stratum corneum lipids: characterization and regional variations. *J. Lipid Res.* **1983**, *24*, 120-130.
18. Stryer, L. *Biochemistry*; W. H. Freeman: New York, 1988.
19. Michaels, A. S.; Chandrasekaran, S. K.; Shaw, J. E. Drug permeation through human skin: theory and in vitro experimental measurement. *AIChE J.* **1975**, *21*, 985-996.
20. Nemanic, M. K.; Elias, P. M. In situ precipitation: a novel cytochemical technique for visualization of permeability pathways in mammalian stratum corneum. *J. Histochem. Cytochem.* **1980**, *28*, 573-578.
21. Boddé, H. E.; van den Brink, I.; Koerten, H. K.; de Haan, F. H. N. Visualization of in vitro percutaneous penetration of mercuric chloride; transport through intercellular space versus cellular uptake through desmosomes. *J. Control. Release* **1991**, *15*, 227-236.
22. Edwards, D. A.; Langer, R. A linear theory of transdermal transport phenomena. *J. Pharm. Sci.* **1994**, *83*, 1315-1334.
23. Grimnes, S. Pathways of ionic flow through human skin in vivo. *Acta Derm. Venereol. (Stockh)* **1984**, *64*, 93-98.

24 Burnette, R. R.; Ongpipattanakul, B. Characterization of the pore transport properties and tissue alteration of excised human skin during iontophoresis. *J. Pharm. Sci.* **1988**, *77*, 132-137.
25 Cullander, C.; Guy, R. H. Visualization of iontophoretic pathways with confocal microscopy and the vibrating probe electrode. *Solid State Ionics* **1992**, *53-56*, 197-206.
26 Cullander, C. What are the pathways of iontophoretic current flow through mammalian skin? *Adv. Drug Deliv. Rev.* **1992**, *9*, 119-135.
27 Scott, E. R.; Laplaza, A. I.; White, H. S.; Phipps, J. B. Transport of ionic species in skin: contribution of pores to the overall skin conductance. *Pharm. Res.* **1993**, *10*, 1699-1709.
28 Kasting, G. B.; Smith, R. L.; Cooper, E. R. Effect of lipid solubility and molecular size on percutaneous absorption. In *Skin Pharmacokinetics*; Shroot, B.; Schaefer, H., Eds.; Karger: Basel, 1987; pp. 138-153.
29 Tojo, K. Random brick model for drug transport across stratum corneum. *J. Pharm. Sci.* **1987**, *76*, 889 -.
30 Anderson, B. D.; Raykar, P. V. Solute structure-permeability relationships in human stratum corneum. *J. Invest. Dermatol.* **1989**, *93*, 280-286.
31 Potts, R. O.; Guy, R. H. Predicting skin permeability. *Pharm. Res.* **1992**, *9*, 663-669.
32 Harrison, J. E.; Watkinson, A. C.; Green, D. M.; Hadgraft, J.; Brain, K. The relative effect of azone and transcutol on permeant diffusivity and solubility in human stratum corneum. *Pharm. Res.* **1996**, *13*, 542-546.
33 Scheuplein, R. J.; Blank, I. H. Permeability of the skin. *Physiol. Rev.* **1971**, *51*, 702-747.
34 Elias, P. M. Epidermal lipids, membranes, and keratinization. *Int. J. Dermatol.* **1981**, *20*, 1.
35 Flynn, G. L. Mechanism of percutaneous absorption from physicochemical evidence. In *Percutaneous Absorption*; Bronaugh, R. L.; Maibach, H. I., Eds.; Marcel Dekker: New York, 1989; pp. 27-51.
36 Yamashita, F.; Bando, H.; Koyama, Y.; Kitagawa, S.; Takakura, Y.; Hashida, M. In vivo and in vitro analysis of skin penetration enhancement based on a two-layer diffusion model with polar and nonpolar routes in the stratum corneum. *Pharm. Res.* **1994**, *11*, 185-191.
37 Yoneto, K.; Li, S. K.; Ghanem, A. H.; Crommelin, D. J.; Higuchi, W. I. A mechanistic study of the effects of 1-alkyl-2-pyrrolidones on bilayer permeability of stratum corneum lipid liposomes: a comparison with hairless mouse skin studies. *J. Pharm Sci.* **1995**, *84*, 853-861.
38 Illel, B.; Schaefer, H.; Wepierre, J.; Doucet, O. Follicles play an important role in percutaneous absorption. *J. Pharm. Sci.* **1991**, *80*, 424-427.
39 Hueber, F.; Wepierre, J.; Schaefer, H. Role of transepidermal and transfollicular routes in percutaneous absorption of hydrocortisone an d testosterone: in vivo study in the hairless rat. *Skin Pharmacol.* **1992**, *5*, 99-107.
40 Santus, G. C.; Baker, R. W. Transdermal enhancer patent literature. *J. Control. Release* **1993**, *25*, 1-20.
41 Walker, R. B.; Smith, E. W. The role of percutaneous penetration enhancers. *Adv. Drug Deliv. Reviews* **1996**, *18*, 295-301.
42 Kastin, A. J.; Arimura, A.; Schally, A. V. Topical absorption of polypeptides with dimethylsufloxide. *Arch. Derm.* **1966**, *93*, 471.
43 Kazim, M.; Weber, C.; Strausberg, L.; LaForet, G.; Nicholson, J.; Reemtsma, K. Treatment of diabetes in mice with topical aplication of insulin to the skin. *Diabetes* **1984**, *33*, 181A.

44 Coapman, S. D.; Lichtin, J. L.; Sakr, A.; Schlitz, J. R. Studies of the penetration of native collagen, collagen alpha chains and collagen cyanogen bromide peptides through hairless mouse skin in vitro. *J. Soc. Cosmet. Chem.* **1988**, *39*, 275.
45 Banerjee, P. S.; Ritschel, W. A. Transdermal permeation of vasopressin. I. Influence of pH, concentration, shaving and surfactant on in vitro permeation. *Int. J. Pharm.* **1989**, *49*, 189-197.
46 Banerjee, P. S.; Ritschel, W. A. Transdermal peermeation of vasopressin. II. Influence of Azone on in vitro and in vivo permeation. *Int. J. Pharm.* **1989**, *49*, 199-204.
47 Boddé, H. E.; Verhoef, J. C.; Ponec, M. Transdermal peptide delivery. *Biochem. Soc. Trans.* **1989**, *17*, 943-945.
48 Mazer, N. A. Pharmacological and biophysical considerations in polypeptide delivery: input functions and routes of administration. *J. Pharm. Sci.* **1989**, *78*, 885-887.
49 Choi, H. K.; Flynn, G. L.; Amidon, G. L. Transdermal delivery of bioactive peptides: the effect on n-decyl methyl sulfoxide, pH, and inhibitors on enkaphalin metabolism and transport. *Pharm. Res.* **1990**, *7*, 1099 -.
50 Hoogstraate, A. J.; Verhoef, J.; Brussee, J.; Ijzerman, A. P.; Spies, F.; Boddé, H. E. Kinetics, ultrastructural aspects and molecular modelling of transdermal peptide flux enhancement by N-alkulazacycloheptanones. *Int. J. Pharm* **1991**, *76*, 37.
51 Braun-Falco, O.; Korting, H. C.; Maibach, H. I., Eds. *Liposome Dermatics*; Springer: Berlin, 1992.
52 Gregoriadis, G., Ed. *Liposome Technology*; CRC Press: Boca Raton, FL, 1992.
53 Schmid, M.-H.; Korting, H. C. Liposomes: a drug carrier system for topical treatment in dermatology. *Crit. Rev. Therap. Drug Carrier Systems* **1994**, *11*, 153-157.
54 Schreier, H.; Bouwstra, J. Liposomes and niosomes as topical drug carriers: dermal and transdermal drug delivery. *J. Control. Release* **1994**, *30*, 1-15.
55 Lasch, J.; Laub, R.; Wohlrab, W. How deep do intact liposomes penetrate into human skin? *J. Control. Release* **1991**, *18*, 55-58.
56 Schaller, M.; Korting, H. C. Interaction of liposomes with human skin: the role of the stratum corneum. *Adv. Drug Deliv. Reviews* **1996**, *18*, 303-309.
57 Bouwstra, J. A.; Hofland, H. E. L.; Spies, F.; Gooris, B. S.; Junginger, H. E. Changes in the structure of the human stratum corneum induced by liposomes. In *Liposome Dermatics*; Braun-Falco, O.; Korting, H. C.; Maibach, H. I., Eds.; Springer: Berlin, 1992; pp. 121-136.
58 Touitou, E.; Junginger, H. E.; Weiner, N. D.; Nagai, T.; Mezei, M. Liposomes as carriers for topical and transdermal delivery. *J. Pharm. Sci.* **1994**, *83*, 1189-1203.
59 Lauer, A. C.; Lieb, L. M.; Ramachandran, C.; Flynn, G. L.; Weiner, N. D. Transfullicular drug delivery. *Pharm. Res.* **1995**, *12*, 179-186.
60 Cevc, G.; Blume, G. Lipid vesicles penetrate into intact skin owing to the transdermal osmotic gradients and hydration force. *Biochim. Biophys. Acta* **1992**, *1104*, 226-232.
61 Cevc, G.; Blume, G.; Schätzlein, A.; Gebauer, D.; Paul, A. The skin: a pathway for systemic treatment with patches and lipid-based agent carriers. *Adv. Drug Deliv. Rev.* **1996**, *18*, 349-378.
62 Egbaria, K.; Ramachandran, C.; Weiner, N. Topical application of liposomally entrapped cyclosporin evaluated by in vitro duffusion studies with human skin. *Skin Pharmacol.* **1991**, *4*, 21-28.

63 Yarosh, D.; Alas, L.; Yee, V.; Oberyszyn, A.; Kibitel, J.; Mitchell, D.; Rosenstein, R.; Spinowitz, A.; Citron, M. Pyrimidine dimer removal enhanced by DNA repair liposomes reduces the incidence of UV skin cancer in mice. *Cancer Res.* **1992**, *52*, 4227-4231.

64 Masini, V.; Bonte, F.; Meybeck, A.; Wepierre, J. Cutaneous bioavailability in hairless rats of tretinoin in liposomes or gel. *J. Pharm. Sci.* **1993**, *82*, 17-21.

65 Niemiec, S.; Ramachandran, C.; Weiner, N. Influence of nonionic liposomal composition on topical delivery of peptide drugs into pilosebaceous units: an in vivo study using the hamster ear model. *Pharm. Res.* **1995**, *12*, 1184-1188.

66 Yarosh, D.; Bucana, C.; Cox, P.; Alas, L.; Kibitel, J.; Kripke, M. Localization of liposomes containing a DNA repair enzyme in murine skin. *J. Invest. Dermatol.* **1994**, *103*, 461-468.

67 Du Plessis, J.; Egbaria, K.; Rachmadnran, C.; Weiner, N. Topical delivery of liposomally encapsulated gamma-interferon. *Antiviral Res.* **1992**, *18*, 259-265.

68 Li, L.; Lishko, V.; Hoffman, R. M. Liposomes can specifically target entrapped melanin to hair follicles in histocultured skin. *In Vitro Cell. Dev. Biol.* **1993**, *29A*, 192-194.

69 Miyachi, Y.; Imamura, S.; Niwa, Y. Decreased skin superoxide dismutase activity by a single exposure of ultraviolet radiation is reduced by liposomal superoxide dismutase pretreatment. *J. Invest. Dermatol.* **1987**, *89*, 111-112.

70 Balsari, A. L.; Morelli, D.; Menard, S.; Veronesi, U.; Colnaghi, M. I. Protection against doxorubicin-induced alopecia in rats by liposome-entrapped monoclonal antibodies. *FASEB J.* **1994**, *8*, 226-230.

71 Li, L.; Lishko, V.; Hoffman, R. M. Liposome targeting of high molecular weight DNA to the hair follicles of histocultured skin: a model for gene therapy of the hair growth processes. *In Vitro Cell. Dev. Biol.* **1993**, *29A*, 258-260.

72 Lieb, L. M.; Ramachandran, C.; Egbaria, K.; Weiner, N. Topical delivery enhancement with multilamellar liposomes via the pilosebaceous route: I. In vitro evaluation using fluorescent techniques with the hamster ear model. *J. Invest. Dermatol.* **1992**, *99*, 108-113.

73 Planas, M. E.; Gonzalez, P.; Rodriguez, L.; Sanchez, S.; Cevc, G. Noninvasive percutaneous induction of topical analgesia by a new type of drug carrier, and prolongation of the local pain insensitivity by anesthetic liposomes. *Anesth. Analg.* **1992**, *75*, 615-621.

74 Cevc, G. Dermal insulin. In *Frontiers in Insulin Pharmacology*; Berger, M.; Gries, A., Eds.; Stuttgart: Georg Thieme, 1993; pp. 161-169.

75 Paul, A.; Cevc, G.; Bachhawat, B. K. Transdermal immunization with large proteins by means of ultradeformable drug carriers. *Eur. J. Immunol.* **1995**, *25*, 3521-3524.

76 Bockris, J. O.; Reddy, A. K. N. *Modern Electrochemistry*; Plenum Press: New York, 1970.

77 Pikal, M. J. The role of electroosmotic flow in transdermal iontophoresis. *Adv. Drug Deliv. Rev.* **1992**, *9*, 201-237.

78 Srinivasan, V.; Sims, S. M.; Higuchi, W. I.; Behl, C. R.; Malick, A. W.; Pons, S. Iontophoretic transport of drugs: a constant voltage approach. In *Pulsed and Self-Regulated Drug Delivery*; Kost, J. Ed.; CRC Press: Boca Raton, FL, 1990; pp. 66-89.

79 Kasting, G. B. Theoretical models for iontophoretic delivery. *Adv. Drug Deliv. Rev.* **1992**, *9*, 177-199.

80 Potts, R. O.; Guy, R. H.; Francoeur, M. L. Routes of ionic permeability through mammalian skin. *Solid State Ionics* **1992**, *53-56*, 165-169.

81 Monteiro-Riviere, N. A.; Inman, A. O.; Riviere, J. E. Identification of the pathway of iontophoretic drug delivery: light and ultrastructural studies using mercuric chloride in pigs. *Pharm. Res.* **1994**, *11*, 251-256.
82 Stephens, W. G. S. The current-voltage relationship in human skin. *Med. Electron. Biol. Eng.* **1963**, *1*, 389-399.
83 Kasting, G. B.; Bowman, L. A. DC electrical properties of frozen, excised human skin. *Pharm. Res.* **1990**, *7*, 134-143.
84 Pikal, M. J.; Shah, S. Transport mechanisms in iontophoresis. III. An experimental study of the contributions of electroosmotic flow and permeability change in transport of low and high molecular weight solutes. *Pharm. Res.* **1990**, *7*, 222-229.
85 Dinh, S. M.; Luo, C.-W.; Berner, B. Upper and lower limits of human skin electrical resistance in iontophoresis. *AIChE J.* **1993**, *39*, 2011-2018.
86 Oh, S. Y.; Leung, L.; Bommannan, D.; Guy, R. H.; Potts, R. O. Effect of current, ionic strength and temperature on the electrical properties of skin. *J. Control. Release* **1993**, *27*, 115-125.
87 Inada, H.; Ghanem, A.-H.; Higuchi, W. I. Studies on the effects of applied voltage and duration on human epidermal membrane alteration/recovery and the resultant effects upon iontophoresis. *Pharm. Res.* **1994**, *11*, 687-697.
88 Prausnitz, M. R. The effects of electric current applied to the skin: a review for transdermal drug delivery. *Adv. Drug Deliv. Rev.* **1996**, *18*, 395-425.
89 Sloan, J. B.; Soltani, K. Iontophoresis in dermatology. *J. Am. Acad. Dermatol.* **1986**, *15*, 671-684.
90 Banga, A. K.; Chien, Y. W. Iontophoretic delivery of drugs: fundamentals, developments and biomedical applications. *J. Control. Release* **1988**, *7*, 1-24.
91 Singh, J.; Roberts, M. S. Transdermal delivery of drugs by iontophoresis: a review. *Drug Design and Delivery* **1989**, *4*, 1-12.
92 Ledger, P. W. Skin biological issues in electrically enhanced transdermal delivery. *Adv. Drug Deliv. Rev.* **1992**, *9*, 289-307.
93 Warwick, W. J.; Huang, N. N.; Waring, W. W.; Cherian, A. G.; Brown, I.; Stejskal-Lorenz, E.; Yeung, W. H.; Duhon, G.; Hill, J. G.; Strominger, D. Evaluation of a cystic fibrosis screening system incorporating a miniature sweat stimulator and disposable chloride sensor. *Clin. Chem.* **1986**, *32*, 850-853.
94 Hölzle, E.; Alberti, N. Long-term efficacy and side effects of tap water iontophoresis of palmoplantar hyperhidrosis — the usefulness of home therapy. *Dermatologica* **1987**, *175*, 126-135.
95 Gangarosa, L. P. *Iontophoresis in Dental Practice*; Quintessence Publishing: Chicago, 1983.
96 Chien, Y. W.; Lelawongs, P.; Siddiqui, O.; Sun, Y.; Shi, W. M. Facilitated transdermal delivery of therapeutic peptides and proteins by iontophoretic delivery devices. *J. Control. Release* **1990**, *13*, 263-278.
97 Lelawongs, P.; Liu, J.-C.; Siddiqui, O.; Chien, Y. W. Transdermal iontophoretic delivery of arginine-vasopressin (I): Physicochemical considerations. *Int. J. Pharm* **1989**, *56*, 13-22.
98 Miller, L. L.; Kolaskie, C. J.; Smith, G. A.; Rivier, J. Transdermal iontophoresis of gonadotropin-releasing hormone (LHRH) and two analogues. *J. Pharm. Sci.* **1989**, *79*, 490-493.
99 Heit, M. C.; Monteiro-Riviere, N. A.; Jayes, F. L.; Riviere, J. E. Transdermal iontophoretic delivery of luteinizing hormone releasing hormone (LHRH): effect of repeated administration. *Pharm. Res.* **1994**, *11*, 1000-1003.
100 Thysman, S.; Hanchard, C.; Preat, V. Human calcitonin delivery in rats by iontophoresis. *J. Pharm. Pharmacol.* **1994**, *46*, 725-730.

101 Kumar, S.; Char, H.; Patel, S.; Piemontese, D.; Malick, A. W.; Iqbal, K.; Neugroschel, E.; Behl, C. R. In vivo transdermal iontophoretic delivery of growth hormone releasing factor GRF (1-44) in hairless guinea pigs. *J. Control. Release* **1992**, *18*, 213-220.
102 Stephen, R. L.; Petlenz, T. J.; Jacobsen, S. C. Potential novel methods for insulin administration: I. Iontophoresis. *Biomed. Biochim. Acta* **1984**, *43*, 553-558.
103 Kari, B. Control of glucose levels in alloxan-diabetic rabbits by iontophoresis of insulin. *Diabetes* **1986**, *35*, 217.
104 Chien, Y. W.; Siddiqui, O.; Sun, Y.; Shi, W. M.; Liu, J. C. Transdermal iontophoretic delivery of therapeutic peptides/proteins. *Ann. N. Y. Acad. Sci.* **1987**, *507*, 32-51.
105 Meyer, B. R.; Katzeff, H. L.; Eschbach, J. C.; Trimmer, J.; Zacharias, S. B.; Rosen, S.; Sibalis, D. Transdermal delivery of human insulin to albino rabbits using electrical current. *Am. J. Med. Sci.* **1989**, *297*, 321-325.
106 Meyer, B. R.; Kreis, W.; Eschbach, J.; O'Mara, V.; Rosen, S.; Sibalis, D. Successful transdermal administration of therapeutic doses of a polypeptide to normal human volunteers. *Clin. Pharmacol. Ther.* **1988**, *44*, 607-612.
107 Meyer, B. R.; Kreis, W.; Eschbach, J.; O'Mara, V.; Rosen, S.; Sibalis, D. Transdermal versus subcutaneous leuprolide: a comparison of acute pharmacodynamic effect. *Clin. Pharmacol. Ther.* **1990**, *48*, 340-345.
108 Srinivasan, V.; Higuchi, W. I.; Sims, S. M.; Ghanem, A. H.; Behl, C. R. Transdermal iontophoretic drug delivery: mechanistic analysis and application to polypeptide delivery. *J. Pharm. Sci.* **1989**, *78*, 370-375.
109 Srinivasan, V.; Su, M.-H.; Higuchi, W. I.; Behl, C. R. Iontophoresis of polypeptides: effect of ethanol pretreatment of human skin. *J. Pharm. Sci.* **1990**, *79*, 588-591.
110 Prausnitz, M. R.; Bose, V. G.; Langer, R.; Weaver, J. C. Transdermal drug delivery by electroporation. *Proceed. Intern. Symp. Control. Rel. Bioact. Mater.* **1992**, *19*, 232-233.
111 Prausnitz, M. R.; Bose, V. G.; Langer, R.; Weaver, J. C. Electroporation of mammalian skin: a mechanism to enhance transdermal drug delivery. *Proc. Natl. Acad. Sci. USA* **1993**, *90*, 10504-10508.
112 Chang, D. C.; Chassy, B. M.; Saunders, J. A.; Sowers, A. E., Eds. *Guide to Electroporation and Electrofusion*; Academic Press: New York, 1992.
113 Orlowski, S.; Mir, L. M. Cell electropermeabilization: a new tool for biochemical and pharmacological studies. *Biochim. Biophys. Acta* **1993**, *1154*, 51-63.
114 Weaver, J. C. Electroporation: a general phenomenon for manipulating cells and tissues. *J. Cell. Biochem.* **1993**, *51*, 426-435.
115 Chizmadzhev, Y. A.; Zarnytsin, V. G.; Weaver, J. C.; Potts, R. O. Mechanism of electroinduced ionic species transport through a multilamellar lipid system. *Biophys. J.* **1995**, *68*, 749-765.
116 Edwards, D. A.; Prausnitz, M. R.; Langer, R.; Weaver, J. C. Analysis of enhanced transdermal transport by skin electroporation. *J. Control. Release* **1995**, *34*, 211-221.
117 Pliquett, U.; Langer, R.; Weaver, J. C. Changes in the passive electrical properties of human stratum corneum due to electroporation. *Biochim. Biophys. Acta* **1995**, *1239*, 111-121.
118 Prausnitz, M. R.; Lee, C. S.; Liu, C. H.; Pang, J. C.; Singh, T.-P.; Langer, R.; Weaver, J. C. Transdermal transport efficiency during skin electroporation and iontophoresis. *J. Control. Release* **1996**, *38*, 205-217.

119 Bose, V. G. *Electrical Characterization of Electroporation of Human Stratum Corneum*; PhD Thesis, Massachusetts Institute of Technology: Cambridge, MA, 1994.
120 Burnette, R. R.; Bagniefski, T. M. Influence of constant current iontophoresis on the impedance and passive Na^+ permeability of excised nude mouse skin. *J. Pharm. Sci.* **1988**, *77*, 492-497.
121 Kalia, Y. N.; Guy, R. H. The electrical characteristics of human skin in vivo. *Pharm. Res.* **1995**, *12*, 1605-1613.
122 Bommannan, D.; Leung, L.; Tamada, J.; Sharifi, J.; Abraham, W.; Potts, R. Transdermal delivery of luteinizing hormone releasing hormone: comparison between electroporation and iontophoresis in vitro. *Proceed. Intern. Symp. Control. Rel. Bioact. Mater.* **1993**, *20*, 97-98.
123 Prausnitz, M. R.; Seddick, D. S.; Kon, A. A.; Bose, V. G.; Frankenburg, S.; Klaus, S. N.; Langer, R.; Weaver, J. C. Methods for in vivo tissue electroporation using surface electrodes. *Drug Delivery* **1993**, *1*, 125-131.
124 Vanbever, R.; Lecouturier, N.; Préat, V. Transdermal delivery of metoprolol by electroporation. *Pharm. Res.* **1994**, *11*, 1657-1662.
125 Hofmann, G. A.; Rustrum, W. V.; Suder, K. S. Electro-incorporation of microcarriers as a method for the transdermal delivery of large molecules. *Bioelectrochem. Bioenerg.* **1995**, *38*, 209-222.
126 Prausnitz, M. R.; Edelman, E. R.; Gimm, J. A.; Langer, R.; Weaver, J. C. Transdermal delivery of heparin by skin electroporation. *Bio/Technology* **1995**, *13*, 1205-1209.
127 Prausnitz, M. R.; Bose, V. G.; Langer, R.; Weaver, J. C. Electroporation. In *Percutaneous Penetration Enhancers*; Smith, E. W.; Maibach, H. I., Eds.; CRC Press: Boca Raton, FL, 1995; pp. 393-405.
128 Vanbever, R.; Preat, V. Factors affecting transdermal delivery of metoprolol by electroporation. *Bioelectrochem. Bioenerget.* **1995**, *38*, 223-228.
129 Zewert, T. E.; Pliquett, U. F.; Langer, R.; Weaver, J. C. Transdermal transport of DNA antisense oligonucleotides by electroporation. *Biochem. Biophys. Res. Com.* **1995**, *212*, 286-292.
130 Pliquett, U.; Weaver, J. C. Electroporation of human skin: simultaneous measurement of changes in the transport of two fluorescent molecules and in the passive electrical properties. *Bioelectrochem. Bioenerget.* **1996**, *39*, 1-12.
131 Regnier, V.; Le Doan, T.; Laurent, M.; Préat, V. Cutaneous delivery of oligonucleotides by electroporation. In *Therapeutic Oligonucleotide Regulation of Gene Expression by Targeting Nucleic Ccids: Chemistry, Biology, Pharmacology and Biological Applications*: Rome, Italy, 9-12 June 1996.
132 Pliquett, U.; Weaver, J. C. Transport of a charged molecule across the human epidermis due to electroporation. *J. Control. Release*, in press.
133 Prausnitz, M. R. Electroporation. In *Electronically Controlled Drug Delivery*; Berner, B.; Dinh, S. M., Eds.; CRC Press: Boca Raton, FL, in press.
134 Prausnitz, M. R. Do high-voltage pulses cause changes in skin structure? *J. Control. Release*, in press.
135 Vanbever, R.; Le Boulengé, E.; Préat, V. Transdermal delivery of fentanyl by electroporation I. Influence of electrical factors. *Pharm. Res.* **1996**, *13*, 559-565.
136 Vanbever, R.; De Morre, N.; Préat, V. Transdermal delivery of fentanyl by electroporation. II. Mechanisms involved in drug transport. *Pharm. Res.*, in press.
137 Prausnitz, M. R.; Pliquett, U.; Langer, R.; Weaver, J. C. Rapid temporal control of transdermal drug delivery by electroporation. *Pharm. Res.* **1994**, *11*, 1834-1837.

138 Pliquett, U.; Zewart, T. E.; Chen, T.; Langer, R.; Weaver, J. C. Imaging of fluorescent molecule and small ion transport through human stratum corneum during high voltage pulsing: localized transport regions are involved. *Biophys. Chem.* **1996**, *58*, 185-204.
139 Prausnitz, M. R.; Gimm, J. A.; Guy, R. H.; Langer, R.; Weaver, J. C.; Cullander, C. Imaging of transport pathways across human stratum corneum during high-voltage and low-voltage electrical exposures. *J. Pharm. Sci.*, in press.
140 Riviere, J. E.; Monteiro-Riviere, N. A.; Rogers, R. A.; Bommannan, D.; Tamada, J. A.; Potts, R. O. Pulsatile transdermal delivery of LHRH using electroporation: drug delivery and skin toxicology. *J. Control. Release* **1995**, *36*, 229-233.
141 Bommannan, D.; Tamada, J.; Leung, L.; Potts, R. O. Effect of electroporation on transdermal iontophoretic delivery of luteinizing hormone releasing hormone (LHRH) in vitro. *Pharm. Res.* **1994**, *11*, 1809-1814.
142 Tamada, J.; Sharifi, J.; Bommannan, D. B.; Leung, L.; Azimi, N.; Abraham, W.; Potts, R. Effect of electroporation on the iontophoretic delivery of peptides in vitro. *Pharm. Res.* **1993**, *10*, S257.
143 Mitragotri, S.; Blankschtein, D.; Langer, R. Ultrasound-mediated transdermal protein delivery. *Science* **1995**, *269*, 850-853.
144 Suslick, K. S. *Ultrasound: Its Chemical, Physical, and Biological Effects*; VCH: Deerfield Beach, FL, 1988.
145 Leighton, T. G. *The Acoustic Bubble*; Academic Press: London, 1994.
146 Stewart, H. F.; Stratmeyer, M. E. *An Overview of Ultrasound: Theory, Measurement, Medical Applications, and Biological Effects (FDA 82-8190)*; U.S. Department of Health and Human Services: Rockville, MD, 1983.
147 Williams, A. R. *Ultrasound: Biological Effects and Potential Hazards*; Academic Press: New York, 1983.
148 Nyborg, W. L.; Ziskin, M. C. *Biological Effects of Ultrasound*; Churchill Livingstone: New York, 1985.
149 *Exposure Criteria for Medical Diagnostic Ultrasound: I. Criteria Based on Thermal Mechanisms (NCRP Report No. 113)*; National Council on Radiation Protection and Measurements: Bethesda, MD, 1992.
150 Barnett, S. B.; ter Haar, G. R.; Ziskin, M. C.; Nyborg, W. L.; Maeda, K.; Bang, J. Current status of research on biophysical effects of ultrasound. **1994**, *20*, 205-218.
151 Antich, T. J. Phonophoresis: the principles of the ultrasonic driving force and efficacy in treatment of common orthopaedic diagnoses. *J. Orthop. Sports Phys. Ther.* **1982**, *4*, 99-102.
152 Kost, J.; Levy, D.; Langer, R. Ultrasound as a transdermal enhancer. In *Percutaneous Absorption. Mechanisms - Methodology - Drug Delivery*; Bronaugh, R. L.; Maibach, H. I., Eds.; Marcel Dekker: New York, 1989; pp. 595-601.
153 Tyle, P.; Agrawala, P. Drug delivery by phonophoresis. *Pharm. Res.* **1989**, *6*, 355-361.
154 Kost, J.; Langer, R. Ultrasound-mediated transdermal drug delivery. In *Topical Drug Bioavailability, Bioequivalence, and Penetration*; Shah, V. P.; Maibach, H. I., Eds.; Plenum Press: New York, 1993; pp. 91-104.
155 McElnay, J. C.; Benson, H. A.; Hadgraft, J.; Murphy, T. M. The use of ultrasound in skin penetration enhancement. In *Pharmaceutical Skin Penetration Enhancement*; Walters, K. A.; Hadgraft, J., Eds.; Marcel Dekker: New York, 1993; pp. 293-.

156 Camel, E. Ultrasound. In *Percutaneous Penetration Enhancers*; Smith, E. W.; Maibach, H. I., Eds.; CRC Press: Boca Raton, FL, 1995; pp. 369-382.
157 Flynn, H. G.; Church, C. Transient pulsations of small gas bubbles in water. *J. Acoust. Soc. Am.* **1988**, *84*, 985-998.
158 Crum, L. A.; Roy, R. A.; Dinno, M. A.; Church, C. C.; Apfel, R. E.; Holland, C. K.; Madanshetty, S. I. Acoustic cavitation produced by microsecond pulses of ultrasound: a discussion of some selected results. *J. Acoust. Soc. Am.* **1992**, *91*, 1113-1119.
159 Holmes, S. A.; Whitfield, H. N. The current status of lithotripsy. *Br. J. Urol.* **1991**, *68*, 337-344.
160 Coleman, A. J.; Saunders, J. E. A review of the physical properties and biological effects of the high amplitude acoustic field used in extracorporeal lithotripsy. *Ultrasonics* **1993**, *31*, 75-89.
161 Sinisterra, J. Application of ultrasound to biotechnology: an overview. *Ultrasonics* **1992**, *30*, 180-185.
162 Fellinger, K.; Schmid, J. *Klinik und Therapies des Chromischen Gelenkrheumatismus*; Maudrich: Vienna, Austria, 1954.
163 Griffin, J. E.; Echternach, J. L.; Price, R. E.; Touchstone, J. C. Patients treated with ultrasonic driven hydrocortisone and with ultrasound alone. *Phys. Ther.* **1967**, *47*, 594-601.
164 Benson, H. A. E.; McElnay, J. C.; Harland, R.; Hadgraft, R. Influence of ultrasound on the percutaneous absorption of nicotinate esters. *Pharm. Res.* **1991**, *8*, 204-209.
165 Levy, D.; Kost, J.; Meshulam, Y.; Langer, R. Effect of ultrasound on transdermal drug delivery to rats and guinea pigs. *J. Clin. Invest.* **1989**, *83*, 2074-2078.
166 Brucks, R.; Nanavaty, M.; Jung, D.; Siegel, F. The effects of ultrasound on the in vitro penetration of ibuprofen through human epidermis. *Pharm. Res.* **1989**, *6*, 697-701.
167 Bommannan, D.; Okuyama, H.; Stauffer, P.; Guy, R. H. Sonophoresis: I. The use of ultrasound to enhance transdermal drug delivery. *Pharm. Res.* **1991**, *9*, 559-564.
168 Bommannan, D.; Menon, G. K.; Okuyama, H.; Elias, P. M.; Guy, R. H. Sonophoresis: II. Examination of the mechanism(s) of ultrasound-enhancer transdermal drug delivery. *Pharm. Res.* **1991**, *9*, 1043-1047.
169 Miyazaki, S.; Mizuoka, H.; Oda, M.; Takada, M. External control of drug release and penetration: enhancement of the transdermal absorption of indomethacin by ultrasound irradiation. *J. Pharm. Pharmacol.* **1991**, *43*, 115-116.
170 Mitragotri, S.; Edwards, D. A.; Blankschtein, D.; Langer, R. A mechanistic study of ultrasonically-enhanced transdermal drug delivery. *J. Pharm. Sci.* **1995**, *84*, 697-706.
171 Mortimer, A. J.; Trollope, B. J.; Villeneuve, E. J.; Roy, A. Z. Ultrasound-enhanced diffusion through isolated frog skin. *Ultrasonics* **1988**, *26*, 348-351.
172 Tachibana, K.; Tachibana, S. Transdermal delivery of insulin by ultrasonic vibration. *J. Pharm. Pharmocol.* **1991**, *43*, 270-271.
173 McElnay, J. C.; Matthews, M. P.; Harland, R.; McCafferty, D. F. The effect of ultrasound on the percutaneous absorption of lignocaine. *Br. J. Clin. Pharm.* **1985**, *20*, 421-424.
174 Pratzel, H.; Dittrich, P.; Kukovetz, W. Spontaneous and forced cutaneous absorption of indomethacin in pigs and humans. *J. Rheumatol.* **1986**, *13*, 1122-1125.
175 Benson, H. A. E.; McElnay, J. C.; Harland, R. Use of ultrasound to enhance the percutaneous absorption of benzydamine. *Phys. Ther.* **1989**, *69*, 113-118.

176 Mitragotri, S. *Ultrasound-Enhanced Transdermal Drug Delivery: Mechanisms and Application*; PhD Thesis, Massachusetts Institute of Technology: Cambridge, MA, 1996.
177 Gaertner, W. Frequency dependence of ultrasonic cavitation. *J. Acoust. Soc. Am.* **1954**, *26*, 977-980.
178 Tachibana, K.; Tachibana, S. Use of ultrasound to enhance the local anesthetic effect of topically applied aqueous lidocaine. *Anesthesiology* **1993**, *78*, 1091-1096.
179 Tachibana, K. Transderaml delivery of insulin to alloxan-diabetic rabbits by ultrasound exposure. *Pharm. Res.* **1992**, *9*, 952-954.
180 Mitragotri, S.; Blankschtein, D.; Langer, R. Transdermal drug delivery using low-frequency sonophoresis. *Pharm. Res.* **1996**, *13*, 411-420.
181 Kost, J.; Pliquett, U.; Mitragotri, S.; Yamamoto, A.; Langer, R.; Weaver, J. Synergistic effect of electric field and ultrasound on transdermal transport. *Pharm. Res.* **1996**, *13*, 633-638.
182 Haak, R.; Gupta, S. K. Pulsatile drug delivery from electrotransport therapeutic systems. In *Pulsatile Drug Delivery, Current Applications and Futute Trends*; Gurny, R; Junginger, H. E.; Peppas, N. A., Eds.; Wissenschaftliche Verlagsgesellschaft: Stuttgart, 1993; pp. 99-112.
183 In reference (*182*), this compound is referred to as protein X, but has been identified as cytochrome c in a personal communication from Ron Haak.

Chapter 9

Amino Acid Derived Polymers for Use in Controlled Delivery Systems of Peptides

S. Brocchini, D. M. Schachter, and J. Kohn[1]

Department of Chemistry, Rutgers, The State University of New Jersey, New Brunswick, NJ 08903

A family of synthetic, tyrosine-derived polyarylates is being studied as a new polymeric matrix system for the controlled release of peptides. The polyarylates are degradable amorphous materials whose backbone structure contains amide bonds. Using the cyclic heptapeptide contained in the platelet integrin glycoprotein IIb/IIIa blocking formulation INTEGRILIN™ as a model, the effect of the polymer structure on peptide miscibility within the polymeric matrix and its effect on the release behavior was investigated. A new co-precipitation technique provided polyarylate-peptide blends that were compression molded without decomposition, deactivation, or detectable aggregation of the peptide. Transparent, pliable compression molded films with high peptide loadings of up to 50% (w/w) were fabricated in this way. In spite of such high loadings, the model peptide was not released from these films over a 30 day exposure to physiological buffer solution at 37 °C. Only when poly(ethylene glycol) (PEG) was added to the formulation was the model peptide released. Release rate was a function of polyarylate structure and the amount and molecular weight of PEG used in the blends. This provided an effective means to modulate the release rate.

The clinical effectiveness of peptide drugs is characterized by contrasting pharmacological properties. While a peptide can be highly efficacious in terms of selectivity and potency, adequate dosing by conventional dosage forms tends to be impaired due to low bioavailability and short half life (*1*). To circumvent these limitations, peptides usually require the custom design of an advanced peptide drug delivery system. Such a delivery system is broadly defined as being either a targeted or a controlled release system. A targeted delivery system directs a drug to a desired biological target. The premise of targeted drug delivery is that the therapeutic index of a drug can be improved when the drug accumu-

[1]Corresponding author

lates selectively in specific tissues, organs, or cell types. In contrast, a controlled release delivery system is designed to release a drug in a predetermined, predictable, and reproducible fashion without directing the drug to a specific biological target.

Controlled release systems encompass a wide range of possible configurations for delivering a therapeutic agent. For example, the use of a mechanical infusion pump is one method to control dosing, however to minimize complications and possible discomfort to the patient, other methods of controlled release may be preferable (2). Notably, transdermal and implantable polymeric delivery systems have been developed and several are marketed. Implantable polymeric controlled release systems offer the advantage that patient compliance is guaranteed and that the dosing process is almost imperceptible to the patient. A commonly used classification scheme based on the physical design of such implantable devices differentiates between matrix and reservoir systems. A matrix system consists of a drug dispersed within a polymer, while a reservoir system consists of a separate drug phase physically confined within a surrounding polymeric phase or membrane.

The use of a degradable (resorbable) polymer would be advantageous in the design of an implantable polymeric controlled release system since the need to retrieve the device after release of the peptide is eliminated. However, the choice of an optimally suitable polymeric matrix for the formulation of implantable peptide release systems is not an easy task. Release of the peptide from a polymeric matrix system is controlled by a combination of drug diffusion through the matrix, imbibation of water into the matrix, chemical degradation of the polymer, physical erosion of the matrix system, and the stability of the peptide within the polymeric matrix. Optimizing this array of parameters for an implantable device designed to deliver a specific peptide is both a scientific and technological challenge.

Available Polymers

Several implantable degradable controlled release systems for peptides have been studied using poly(lactic acid) (PLA), poly(glycolic acid) (PGA) or copolymers thereof (1,3). The main advantage of these polymers is their extensive prior history of clinical use and their general acceptance by scientists and regulatory agencies alike as biocompatible and safe materials. However, these polymers are characterized by relatively high equilibrium water contents which can surpass 20% (by weight). The incorporation of water soluble peptides into the polymeric matrix tends to increase water uptake even further and can lead to substantial swelling accompanied by irregular release profiles.

A number of alternative degradable polymers are available, including polycaprolactone (4), poly(hydroxy butyrate) (5,6), poly(ortho esters) (7-10), polyphosphazenes (11), polyanhydrides (12,13). Synthetic poly(amino acids) (1,14-19) as well as natural biopolymers such as proteins (e.g., gelatin, collagen, albumin) (20) and polysaccharides (e.g., dextran, chitosan, starch, cellulose derivatives) (21)) have also been considered as matrix materials in the design of controlled release systems.

One of the severe limitations in the design of controlled release systems for peptides and proteins is the observation that these moieties tend to lose their biological activity upon storage within the polymeric matrix. Based on the hypothesis that "protein-like" materials may stabilize the peptide drugs dispersed therein, it has been implied that such "protein-like" matrices may be particularly useful for the design of release systems for peptides and proteins. Although poly(amino acids) share many of the structural features of peptides, such materials have been of little practical use because they are relatively non-processible, decompose in the molten state, and swell in moist environments. These limitations have been largely overcome in a recently developed class of amino acid derived polymers known as pseudo-poly(amino acids). Pseudo-poly(amino acids) are derived from α-amino acids but contain non-amide bonds (e.g., carbonate or ester linkages) in their backbone (*22,23*).

A wide range of different pseudo-poly(amino acids) has been prepared (*22*). Of particular interest for the design of peptide drug delivery systems are the tyrosine-derived polyarylates **3** (Figure 1) (*24*). These materials are copolymers of a tyrosine-derived dipeptide **1** and a diacid **2**. This family of polyarylates is being used (i) to explore possible correlations between systematic changes in polymer structure and the peptide-polymer interactions, and (ii) to design novel controlled release systems for selected peptides.

Tyrosine-Derived Polyarylates

Tyrosine-derived polyarylates **3** are amorphous materials which are prepared by the polymerization (Figure 1) of one of four alkyl esters of desaminotyrosyl-tyrosyl dipeptide **1** and one of four aliphatic acyclic diacids **2** to give a family of 16 polymers (Table I). In effect each polyarylate is systematically homologated by the addition of methylene groups to either the pendent chain or polymer backbone. Varying the number of methylene groups in this manner produces small changes in polymer structure resulting in incremental changes in most polymer properties. While thermomechanical properties, surface hydrophobicity, and miscibility with drugs and peptides differ dramatically among individual polyarylates, all members of this family are similar in regard to processibility, rate and mechanism of degradation, and biocompatibility. Thus, it is possible to select specific polyarylates to match closely the requirements for the release of a given peptide, or one can use a range of polyarylates as model polymers to study the correlations between polymer structure and the observed release profiles.

General Considerations. The polyarylate backbone contains peptide-like amide bonds. The hydrogen bond donating and accepting sites associated with the amide bonds present in all polyarylates can be expected to increase the solubility of peptides within the polymer matrix while inhibiting peptide aggregation.

Although the polyarylates possess a number of hydrophilic and possible peptide solubilizing sites, films and pins fabricated from polyarylates absorb very little water (<1-3%) at 37°C up to time periods of 25 weeks (*23-26*)). The use of a polymer matrix with low water uptake is important for the release of peptides because excessive water uptake will contribute to peptide aggregation and possible deactivation. A polymer exhibiting a significant degree of water absorption, especially over a short period of time, probably would not be useful for the controlled release of a peptide over longer periods of time, e.g. ranging from weeks to months.

Table I. Family of 16 polyarylates [a]

Diacid	Alkyl Esters of Tyrosyl-Desaminotyrosyl Dipeptide Diphenol Component			
$HO-\overset{O}{\overset{\|}{C}}-(CH_2)_y-\overset{O}{\overset{\|}{C}}-OH$ **2**	$HO-\langle\bigcirc\rangle-CH_2-CH_2-\overset{O}{\overset{\|}{C}}-NH-\overset{\underset{\|}{C=O}}{\underset{\underset{R}{\overset{\|}{O}}}{CH}}-CH_2-\langle\bigcirc\rangle-OH$ **1**			
	DTE	DTB	DTH	DTO
Succinic (y=2)	poly(DTE succinate)	poly(DTB succinate)	poly(DTH succinate)	poly(DTO succinate)
Adipic (y=4)	poly(DTE adipate)	poly(DTB adipate)	poly(DTH adipate)	poly(DTO adipate)
Suberic (y=6)	poly(DTE suberate)	poly(DTB suberate)	poly(DTH suberate)	poly(DTO suberate)
Sebacic (y=8)	poly(DTE sebacate)	poly(DTB sebacate)	poly(DTH sebacate)	poly(DTO sebacate)

[a]The desaminotyrosyl-tyrosine (DT) alkyl esters are abbreviated DTE, DTB, DTH, and DTO where E=ethyl, B=butyl, H=hexyl, and O=octyl, respectively.

Synthesis. The tyrosyl-desaminotyrosyl dipeptides **1** can be prepared in the laboratory on a 200-300 g scale at high enough purity for polymerization (27). Polymerization at ambient temperature using 2.5-3.0 equivalents of diisopropylcarbodiimide mediated by 0.4 equivalent of 4-dimethylaminopyridine-p-toluenesulfonic acid complex is mild (28) with isolated yields of polymer around 65-75%. Polyarylates prepared in this fashion typically have molecular weights ranging from 80,000-150,000 g/mol with polydispersities between 1.5-1.8 as calculated by GPC relative to polystyrene standards in tetrahydrofuran. Low angle light scattering data confirm that this set of GPC conditions is sufficient for obtaining acceptable estimates of the weight average molecular weights of the polyarylates.

Material Properties. The polyarylates possess graduated and incrementally differentiated properties which vary over a wide range as a function of the number of methylene groups in the pendent chain or polymer backbone. As expected, the glass transition temperature (T_g) decreases and the air-water contact angle increases as methylene groups are added to either the pendent chain or polymer backbone (Tables II-III).

These properties vary differently for pendent chain modifications as compared to backbone modifications. The addition of methylene groups to the polymer backbone causes a greater decrease in the T_g than addition of methylene groups to the pendent chain (Table II). In contrast, the air-water contact angle as a measure of surface hydrophobicity increases more as methylene groups are added to the pendent chain than to the polymer backbone (Table III).

Table II. Glass transition temperatures (°C) of the polyarylates [a]

	DTE	DTB	DTH	DTO
succinate	78	67	53	48
adipate	61	46	38	28
suberate	50	37	27	21
sebacate	43	30	20	13

[a]Values reported at the midpoint of the baseline shift in the DSC thermogram

Table III. Air-water contact angles of the polyarylates [a]

	DTE	DTB	DTH	DTO
succinate	68	74	81	86
adipate	73	77	84	87
suberate	76	81	85	88
sebacate	79	84	90	95

[a]Reported as the average of 10 measurements by the same experimentalist.

Evaluation of the tensile properties indicates that the polyarylates with T_g values above ambient temperature are relatively strong materials comparable to other nonreinforced degradable polymers such as poly(D, L-lactic acid), poly(ortho esters), and poly(hydroxybutyrate) (24). Polyarylates with T_g values below ambient temperature are soft, extremely pliable materials similar to soft silicon rubbers.

Processibility. Polyarylates are processible and tend to exhibit very similar processing characteristics across the 16 tested polymers. For example, all polyarylates are soluble in common volatile solvents such as methylene chloride and tetrahydrofuran in concentrations up to 10-12% (w/v) depending on polymer molecular weight. All 16 polyarylates can be readily solvent cast into clear transparent films using 5-7% (w/v) methylene chloride solutions. Spin coating with 2.5% (w/v) methylene chloride solutions yields clear evenly coated surfaces.

Thermogravimetric analysis indicates that the polyarylates decompose at temperatures ranging from 330-360 °C. This implies a large temperature window between T_g and decomposition which is advantageous for processing into devices such as films and rods which are suitable for implantation. Importantly, the T_g values of the polyarylates are relatively low (13-78 °C). This is critical for fabricating release systems for therapeutic peptides or proteins since these moieties are heat sensitive and require low processing temperatures to minimize peptide deactivation. Most tyrosine-derived polyarylates can be compression molded at temperatures as low as 60 °C, yielding clear transparent films. In addition, rods and pins, ribbons, and continuous fibers were prepared from polyarylates by extrusion. The use of these thermal processing techniques results in molecular weight reductions of less than 10%.

Hydrolytic Degradation. Three polymers, poly(DTE adipate), poly(DTH adipate) and poly(DTO adipate), having different pendent chains but an identical polymer backbone were used to compare the rate of hydrolytic degradation among different polyarylates. Under in vitro conditions (37 °C in pH 7.4 phosphate buffered saline), molecular weight reductions of about 50% (24) were observed for all three polymers over a 26 week period. The absence of a measurable variation in the rate of backbone cleavage was an unexpected result: The three test polymers differed not only in their overall hydrophobicity, but poly(DTE adipate) (T_g = 61 °C) was initially in the glassy state, poly(DTH adipate) (T_g = 38 °C) was at its transition point, and poly(DTO adipate) (T_g = 28 °C) was in the rubbery state at the conditions of the degradation study. In spite of their compositional and morphological differences, all three polyarylates seemed to degrade in vitro at similar rates. This unexpected result is currently under further investigation (Kohn et al., manuscript in preparation).

The weight of the film specimens remained essentially unchanged for the duration of this study with very little absorption of water. Preliminary results from ongoing in vitro and in vivo degradation studies with compression molded films and extruded pins suggest that low water uptake is a general property of the polyarylates. Furthermore, ongoing studies indicate that the rate of polyarylate degradation is identical in vitro and in vivo.

Biocompatibility. Detailed evaluations of the biocompatibility of tyrosine-derived polyarylates are currently in progress. Based on preliminary data obtained from cytotoxicity assays in vitro and subcutaneous implantation studies in rats, polyarylates are not cytotoxic and elicit a mild foreign body response comparable to the response seen for commonly used biomaterials such as medical grade polyethylene or PLA (Kohn et al., manuscript in preparation).

Fabrication of Peptide Loaded Polyarylate Films

A commonly used fabrication technique referred to as "suspension casting" is based on suspending a peptide in powdered form in an organic solution of a polymer followed by solvent casting (29). This results in a device where discrete particles of peptide are dispersed within the polymeric phase. At low loadings, little or no peptide is generally released because the phase-separated microdomains of solid peptide particles are surrounded by polymer with little connectivity between adjacent peptide domains (30). At higher loading, such heterogeneity favors a burst-like release of the peptide.

This technique has been used extensively but with variable success (31). One of the common problems encountered in the formulation of controlled release matrix systems for peptides is the lack of a common solvent for both peptide and polymer: while the synthetic polymers used as matrix materials are usually soluble in organic solvents, peptides are often soluble in aqueous solutions only. Thus, it can be difficult to obtain uniform and fine dispersions of the peptide within the polymeric matrix.

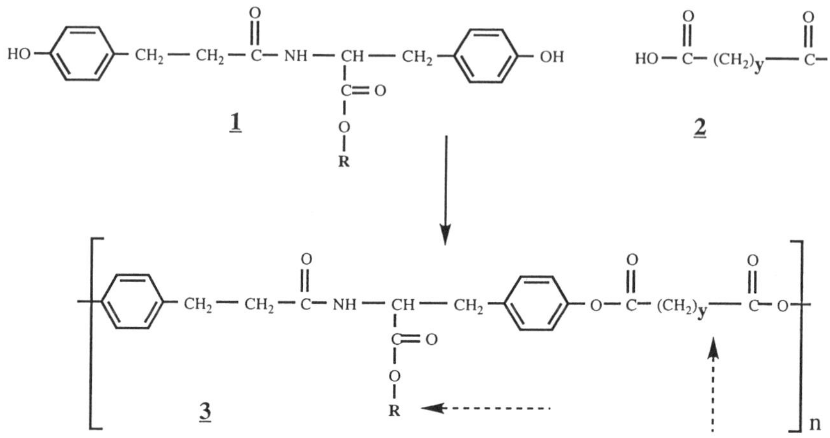

Figure 1: Polyesterification of desaminotyrosyl-tyrosyl dipeptide **1** (R=ethyl, butyl, hexyl, and octyl) and acyclic aliphatic diacids **2** (y=2,4,6,8) to give the polyarylates **3**. Reactions are conducted in methylene chloride at ambient temperature using 2.5-3.0 equivalents of diisopropylcarbodiimide and 0.4 equivalent of 4-dimethylaminopyridine-p-toluenesulfonic acid complex

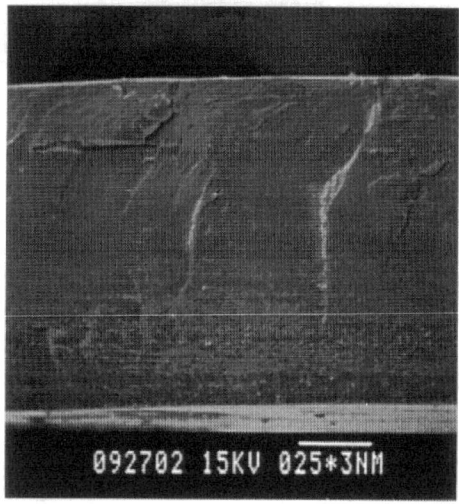

Figure 2: SEM micrograph (150x) of a freeze-fractured film cross-section of a compression molded poly(DTH adipate) film formulated with 20% w/w of a model peptide.

Since the polyarylates can be compression molded into films at relatively low temperatures, a co-precipitation technique was used to more uniformly mix peptide and polymer, thus avoiding some of the shortcomings of the original suspension casting technique (25). Briefly, a peptide is dissolved in a suitable solvent mixture in which both peptide and polyarylate are soluble. This is followed by precipitation into a non-solvent such as ether or hexane in which both peptide and polymer are insoluble. If the precipitation occurs rapidly, an intimately mixed peptide-polyarylate co-precipitate is formed which is dried and then molded.

For example, INTEGRILIN™ (antithrombotic injection) contains a synthetic cyclic heptapeptide which is a highly potent glycoprotein IIb/IIIa antagonist (32). The cyclic heptapeptide could be dissolved in a mixture of methanol, methylene chloride, and the polyarylate to give a homogeneous solution of peptide and polymer. Upon precipitation into cold diethyl ether, a co-precipitate was formed in which no discrete particles of peptide could be detected. After drying in vacuum, the peptide-polymer co-precipitate was an easy-to-handle, solid material.

Using this technique, polyarylates were routinely blended with up to 30% (w/w) of the cyclic heptapeptide. After compression molding of the respective co-precipitates, clear, transparent films were obtained (25). These compression molded film specimens were smooth, pliable, mechanically strong. GPC data indicated no loss in polymer molecular weight due to compression molding. TGA indicated that the films were devoid of residual solvents. No decomposition or irreversible aggregation of the peptide could be detected by a careful, **method-validated** HPLC analysis of the peptide released from the films. Analysis of the morphology of films loaded with 30% (w/w) of the peptide by SEM revealed a smooth, structureless appearance of freeze-fractured film cross-sections. In particular, the films were devoid of detectable domains or discrete peptide particles (Figure 2).

The co-precipitation technique described above is facilitated by the relatively low processing temperature of the polyarylates (50 to 80 °C). In the past, the preparation of controlled release systems for peptides by compression molding, extrusion, or injection molding has not been widely investigated since few polymers were available with sufficiently low processing temperatures.

Study of Peptide Release from Compression Molded Films

Compression molded films formulated from poly(DTE adipate), poly(DTH adipate), and poly(DTO adipate) with loadings ranging from 5-20% (w/w) of the cyclic heptapeptide from INTEGRILIN™ were fabricated to illustrate the influence imparted by small changes in the polyarylate structure (25). A reduction in T_g of a polymer blend is commonly used to assess the miscibility of the blend components. Listed in Table IV are the onset T_g values showing a depression in these values for poly(DTE adipate) and poly(DTH adipate). Interestingly, the Tg values were most depressed at 15% peptide loading for these two polymers. The more hydrophobic polyarylate, poly(DTO adipate) did not exhibit an evident T_g depression indicating non-miscibility of the peptide with the polymer. In spite of the apparent non-miscibility of poly(DTO adipate) and peptide, all the films in the study were transparent, free of visible drug particles, and pliable with excellent handling properties.

Table IV. Glass transition temperatures for compression molded films[a] as a function of peptide loading

Polymer	Glass transition temperature (°C)			
	neat polymer	10% loading	15% loading	20% loading
poly(DTE-adipate)	62.5	44.2	43.2	49.2
poly(DTH-adipate)	35.0	33.3	29.0	33.4
poly(DTO-adipate)	24.0	24.6	24.5	24.4

[a]0.1-0.2 mm thick films molded at 6,000 lbs

Release studies under simulated physiological conditions (37 °C in pH 7.4 phosphate buffer) indicated no significant release (3±2% of total loading) of peptide over a 30 day period from any of the films listed in Table IV, even when the peptide loading was increased to 50% (w/w).

In correspondence with the hydrophobic nature of polyarylates, peptide free control films of poly(DTH adipate) have an equilibrium water content of only 1-2% (w/w). In contrast, water uptake studies of 30% (w/w) peptide loaded poly(DTH adipate) films indicated an equilibrium water content of about 10-12% (w/w) at 37 °C in pH 7.4 phosphate buffer. In spite of the significant uptake of water by peptide-loaded films, the films did not become opaque during the incubation period and (as outlined above) no significant release of peptide occurred.

To explore this phenonmenon further, poly(DTH adipate) film samples were prepared containing a 50% (w/w) loading of peptide. At this high loading, the film was slightly cloudy and less pliable than films at 30% (w/w) peptide loading. The total loading was confirmed to be 50±2% (w/w) by dissolving a weighed sample of film in tetrahydrofuran, extracting the peptide into aqueous solution, followed by quantitative HPLC analysis. Again, a preliminary release study under simulated physiological conditions (37 °C in pH 7.4 phosphate buffer) resulted in only 3-5% release of the loaded peptide over a 5 day period.

The inability to release the entrapped peptide even from heavily loaded films may be related to the intimate blending of polymer and peptide that occurs during the co-precipitation-compression molding fabrication process. In addition, the apparent miscibility of the peptide within the polyarylate matrix may be a contributing factor. Covalent conjugation of the peptide to the polyarylate during film fabrication can be excluded as a possible explanation for the observed entrapment of the peptide within the polymeric matrix since all the peptide can be extracted into water by simply dissolving the polymer in an organic solvent. Furthermore, since peptide extracted from the films was unchanged and biologically active, any irreversible degradation/aggregation within the polymeric matrix is not likely to occur. However, we currently cannot exclude the existence of reversibly aggregated peptide structures within the compression molded films that could contribute to decreased peptide diffusion. Thus, further study is necessary to fully understand the observed entrapment of the cyclic heptapeptide within polyarylate matrices.

Modulated Peptide Release From Polyarylate Films

The apparent permanent entrapment of the cyclic heptapeptide contained in INTEGRILIN™ within the polyarylate matrix provided a unique opportunity for studying ways to modulate peptide release. In an attempt to increase the release of peptide, poly(ethylene glycol) (PEG) was added to the peptide-polyarylate blends. Since PEG is water soluble and structurally unrelated to both the polyarylates and the model peptide, the addition of PEG resulted in phase separation as indicated by the cloudy (translucent) appearance of the resulting films. However, no discrete drug particles were visible and as before, the films were pliable and easy to handle.

In analogy to a recent report (31), the addition of PEG was an effective means to control the release of the cyclic heptapeptide from the polymer matrix. The release profiles of 30% (w/w) peptide loaded poly(DTH adipate) films with PEG (Mn=1000) weight loadings of 7, 10, and 14% (w/w) are shown in Figure 3. Increasing amounts of PEG caused increasing amounts of the cyclic heptapeptide to be released.

As a means to probe the influence of polyarylate structure on peptide release, PEG (Mn=1000) was blended (10% w/w) with constant peptide loading (30% w/w) using four different polyarylates. Release studies with these films (Figure 4) indicated that polyarylate structure did significantly effect peptide release. The data in Figure 4 indicate that small changes in the polymer backbone have a greater effect on the release profile than similar changes in the pendent chain.

Increasing the molecular weight of the PEG from 1000 to 20,000 also influenced release of the cyclic heptapeptide. Typically, an increase in PEG molecular weight caused a decrease in the cumulative peptide release for a given polymer and peptide loading. For example, with a constant loading of cyclic heptapeptide (30% w/w) in poly(DTE succinate) films, peptide release was decreased when PEG of higher molecular weight was used (Figure 5).

Entrapment of the cyclic heptapeptide in the polyarylate films was dramatically decreased by the addition of PEG to the peptide-polymer blend. We currently hypothesize that PEG disrupted the polyarylate-peptide phase. Within the ternary system consisting of polyarylate-peptide-PEG, the rate of peptide release could be effectively controlled by changes in the polymer structure, the amount of PEG present, and/or the PEG molecular weight.

Conclusion

An important observation made during these studies was the significant effect small changes in the polymer structure had on the release of the cyclic heptapeptide. Since many other laboratories use only a limited number of polyesters (in particular, poly(lactic

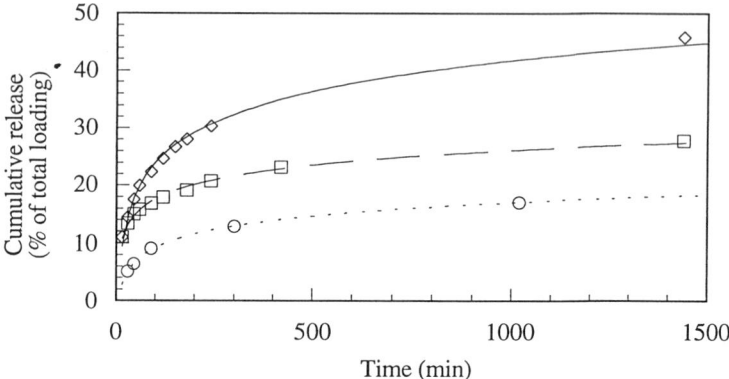

Figure 3: Peptide release from poly(DTH adipate) films at constant peptide load (30% w/w) as a function of the amount of blended PEG (Mn=1000). The release medium was simulated physiological conditions (37 °C in pH 7.4 phosphate buffer). o 7% (w/w) PEG; □ 10% (w/w) PEG; ◇ 14% (w/w) PEG.

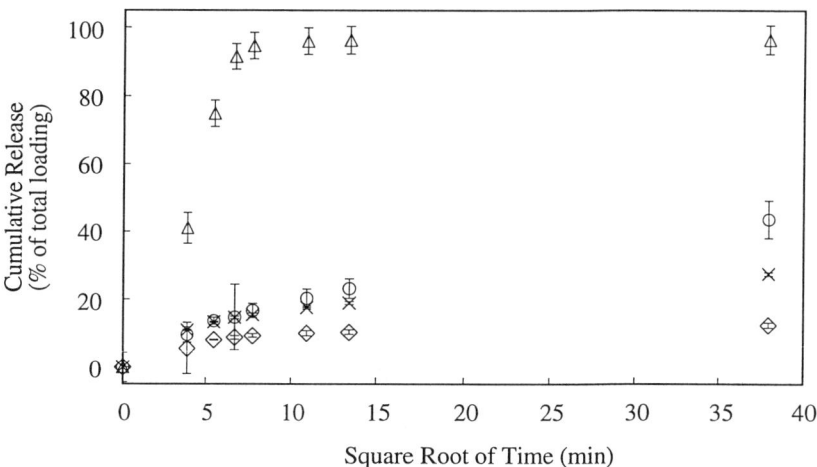

Figure 4: Peptide release profiles (37 °C in pH 7.4 phosphate buffer) as a function of polyarylate structure at constant loading of peptide (30% w/w) and PEG (Mn=1000, 10% w/w). Δ poly(DTE succinate); o poly(DTO succinate); x poly-(DTH adipate); ◇ poly(DTE sebacate).

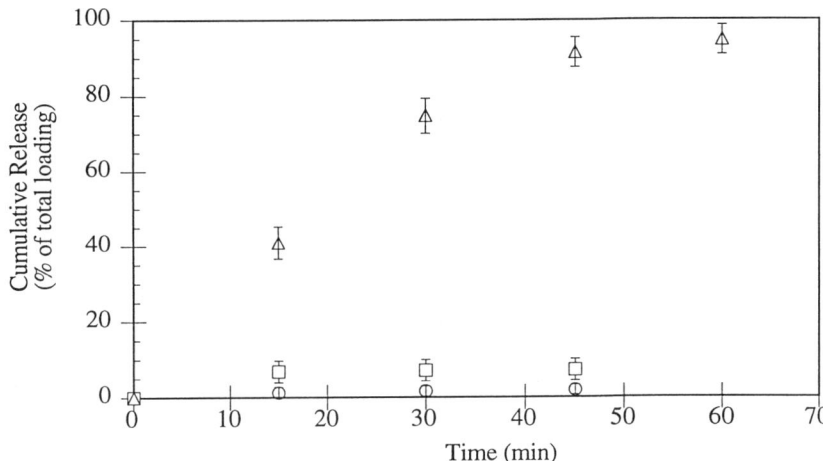

Figure 5: Peptide release (37 °C in pH 7.4 phosphate buffer) from poly(DTE succinate) films as a function of the molecular weight of PEG. All films were prepared in an identical fashion with 30% (w/w) of peptide and 10% (w/w) of PEG. △ PEG Mw=1000; □ PEG Mw=20,000; ○ control film without PEG.

acid), poly(glycolic acid) and copolymers thereof), design parameters relating to the interactions and miscibility between the peptide and the polymeric matrix of release systems are rarely investigated. Here the availability of a "family" of structurally related polymers is facilitating a more systematic approach in our studies. The use of co-precipitation provided uniform blends of peptide and polymer which were compression molded to yield transparent films with excellent handling properties even at high peptide loadings (typically 30% w/w but possible up to 50% w/w). Since the polyarylates can be processed at low temperatures many other devices may potentially be fabricated without decomposition and deactivation of the loaded peptide. Release studies established that under simulated physiological conditions most of the cyclic heptapeptide remained within the polyarylate matrix up to period of thirty days. Peptide entrapment was unexpected because the films were heavily loaded in peptide and imbibed over 10% (w/w) of water. To address this, PEG was added in the co-precipitation process. This gave uniformly mixed ternary blends of polyarylate, peptide, and PEG that upon compression molding produced translucent, pliable films. These films did release the cyclic heptapeptide in correlation to the amount and molecular weight of the added PEG. In addition, peptide release was also dependent on small variations in the polyarylate structure.

Acknowledgements

The authors thank Charles P. du Meé and Joseph Lambing for helpful discussions. INTEGRILIN™ was a gift of COR Therapeutics (South San Francisco, CA) which also supported the work.

Literature Cited

1. Hutchinson, F. G.; Furr, B. J. A. *J. Controlled Release* **1990**, *13*, 279-294.
2. Sinko, P.; Kohn, J. In *Polymeric Delivery Systems: Properties and Applications*; El-Nokaly, M. A., Piatt, D. M. and Charpentier, B. A., Ed.; ACS Symposium Series; Vol. 520, American Chemical Society: Washington, DC, 1993; 18-41.
3. Lewis, D. H. In *Biodegradable Polymers as Drug Delivery Systems*; Chasin, M. and Langer, R., Ed., Marcel Dekker: New York, NY, 1990; 1-41.
4. Pitt, C. G. In *Biodegradable Polymers as Drug Delivery Systems*; Chasin, M. and Langer, R., Ed., Marcel Dekker: New York, NY, 1990; 71-120.
5. Barham, P. J.; Keller, A.; Otun, E. L.; Holmes, P. A. *J. Mater. Sci.* **1984**, *19*, 2781-2794.
6. Miller, N. D.; Williams, D. F. *Biomaterials* **1987**, *8*, 129-137.
7. Heller, J. *J. Control. Rel.* **1985**, *2*, 167-177.
8. Heller, J.; Sparer, R. V.; Zentner, G. M. In *Biodegradable Polymers as Drug Delivery Systems*; Chasin, M. and Langer, R., Ed., Marcel Dekker: New York, NY, 1990; 121-162.
9. Heller, J. *J. Bioact. Compat. Polym.* **1988**, *3 (2)*, 97-105.
10. Heller, J. *Adv. Drug Delivery Rev.* **1993**, *10 (2-3)*, 163-204.
11. Allcock, H. R. In *Biodegradable Polymers as Drug Delivery Systems*; Chasin, M. and Langer, R., Ed., Marcel Dekker: New York, NY, 1990; 163-193.
12. Leong, K. W.; Brott, B. C.; Langer, R. *J. Biomed. Mater. Res.* **1985**, *19*, 941-955.
13. Langer, R. *Chemistry in Britain* **1990**, *26 (3)*, 232-236.
14. Anderson, J. M.; Spilizewski, K. L.; Hiltner, A. In *Biocompatibility of Tissue Analogs*; Williams, D. F., Ed.; Vol. 1, CRC Press Inc.: Boca Raton, 1985; 67-88.
15. Bennett, D. B.; Adams, N. W.; Li, X.; Feijen, J.; Kim, S. W. *J. Bioact. Compat. Polym.* **1988**, *3*, 44-52.
16. Hudecz, F.; Clegg, J. A.; Kajtar, J.; Embleton, M. J.; Pimm, M. V.; Szekerke, M.; Baldwin, R. W. *Bioconj. Chem.* **1993**, *4 (1)*, 25-33.
17. Li, X.; Bennett, D. B.; Adams, N. W.; Kim, S. W. In *Polymeric Drug and Drug Delivery Systems*; Dunn, R. L. and Ottenbrite, R. M., Ed.; ACS Symposium Series; Vol. 469, American Chemical Society: Washington, DC, 1991; 101-116.
18. McCormick-Thomson, L. A.; Duncan, R. *J. Bioact. Biocompat. Polym.* **1989**, *4*, 242-251.
19. Zunino, F.; Savi, G.; Giuliani, F.; Gambetta, R.; Supino, R.; Tinelli, S.; Pezzoni, G. *Eur. J. Cancer Clin. Oncol.* **1984**, *20 (3)*, 421-425.
20. Bogdansky, S. In *Biodegradable Polymers as Drug Delivery Systems*; Chasin, M. and Langer, R., Ed.; Drugs and the Pharmaceutical Sciences; Vol. 45, Marcel Dekker, Inc.: New York, NY, 1990; 231-260.

21. Brode, G. L. In *Cosmetic and Pharmaceutical Applications of Polymers*; Gebelein, C. G., Cheng, T. C. and Yang, V. C., Ed.; Proceedings of an American Chemical Society Symposium on Polymers for Cosmetic and Pharmaceutical Applications held August 1990 in Washington, DC., Plenum Press: New York, NY, 1991; 105-115.
22. Kohn, J. *Trends Polym. Sci.* **1993**, *1 (7)*, 206-212.
23. Kohn, J.; Brocchini, S. In *Polymeric Materials Encyclopedia*; Salamone, J. C., Ed.; Vol. 9, CRC Press: Boca Raton, FL, 1996; 7279-7290.
24. Fiordeliso, J.; Bron, S.; Kohn, J. *J. Biomater. Sci. (Polym. Ed.)* **1994**, *5 (6)*, 497-510.
25. Kohn, J.; Brocchini, S.; Imai, M.; Vyavahare, N. In *22th International Symposium on Controlled Release of Bioactive Materials* [Seattle, WA]; The Controlled Release Society: Deerfield, IL, 1995; 522-523.
26. Schachter, D. M.; Brocchini, S.; Kohn, J. In *Advances in Controlled Delivery* [Baltimore, MD]; Controlled Release Society: Deerfield, IL, 1996; 49-50.
27. Hooper, K. A.; Kohn, J. *J. Bioact. Compat. Polym.* **1995**, *10 (4)*, 327-340.
28. Moore, J. S.; Stupp, S. I. *Macromolecules* **1990**, *23 (1)*, 65-70.
29. Rhine, W. D.; Hsieh, D. S. T.; Langer, R. *J. Pharm. Sci.* **1980**, *69 (3)*, 265-270.
30. Siegel, R. A.; Langer, R. *Pharm. Res.* **1984**, *1*, 2-10.
31. Park, T. G.; Cohen, S.; Langer, R. *Macromolecules* **1992**, *25*, 116-122.
32. Tcheng, J. E.; Harrington, R. A.; Kottke-Marchant, K.; Kleiman, N. S.; Ellis, S. G.; Kereiakes, D. J.; Mick, M. J.; Navetta, F. I.; Smith, J. E.; Worley, S. J.; Miller, J. A.; Joseph, D. M.; Sigmon, K. N.; Kitt, M. M.; Meé, C. P. d.; Califf, R. M.; Topol, E. J. *Circulation* **1995**, *91 (8)*, 2151-2157.

Chapter 10

Analysis of the Solution Behavior of Protein Pharmaceuticals by Laser Light Scattering Photometry

Gay-May Wu, David Hummel, and Alan Herman

Department of Analytical Research and Development, Amgen, Inc., 1840 DeHavilland Drive, 25-2-A, Thousand Oaks, CA 91320

This article presents an overview of the applications of light-scattering to the investigation of protein interactions. Examples include characterization of protein pharmaceuticals (active form of a self-associating protein; composition of a protein conjugate,) detection of soluble aggregate formation (as it relates to the stability and, hence, the formulation of a protein product) and elucidation of protein-ligand interaction.

The biological functions of protein pharmaceuticals depend on the specific interactions of these proteins with other molecules. These interactions require correct conformation and exact stoichiometry of the molecules involved. The primary concern of protein pharmaceutics is the maintenance of the purity, integrity, activity, and stability of the products. As early as sixty five years ago, through the observation of reversible precipitation brought about without chemical changes, the denaturation of a natural protein molecule had been defined as a change in its solubility (1). Subsequently it has been shown that some proteins consist of a defined number of identical or different subunits. This higher order structure is needed for functionality. Determination of the molecular weight and molecular weight averages of protein molecules in solution can yield insight into a number of properties of the protein. These can include the degree of heterogeneity, the formation of soluble aggregates, and the degree and type of association states. Since this type of information is useful in determining physical stability, it can be helpful in selecting protein pharmaceutical formulations. Many physical properties of a protein solution are dependent on molecular weight of the solute, e.g., light-scattering, sedimentation, viscosity and other colligative properties. The methods of quantifying these properties allow the molecular weight to be directly calculated without the need for assumptions concerning the physical or chemical structure of the protein molecules. Here we present some applications of multi-angle laser light scattering in studying the solution behavior of proteins, protein conjugates, and protein complexes.

Theoretical background and instrumentation

The relationship between the amount of scattered light and molecular weight can be expressed as follows (2):

$$(K^*c)/(R_\theta) = [1/(M_w P_\theta)] + 2A_2 c + \qquad [1]$$

where

R_θ = the Rayleigh scattering intensity at angle θ
M_w = the weight average molecular weight
P_θ = the scattering function which accounts for angular dependency in finite-size molecules
c = the concentration of solute molecules in the solution
A_2 = the second virial coefficient which accounts for solvent/solute interactions.

K^* is an optical constant and represents the following:

$$K^* = 4\pi^2 n_0^2 (dn/dc)^2 \lambda_0^{-4} N_A^{-1} \qquad [2]$$

where

n_0 = solvent refractive index at the incident radiation(vacuum) wavelength
dn/dc = refractive index increment as a function of concentration
λ_0 = incident radiation (vacuum) wavelength
N_A = Avogadro's number.

The application of equation [1] is limited to solute particles smaller than the wavelength (molecular dimensions < $\lambda/20$) to eliminate interference of the scattered intensity, and to a two-component systems (so that n_0 remains constant). When P_θ is expanded to the first order equation [1] can be transformed as follows:

$$(K^*c)/(R_\theta) = (1/M_w)[1 + (16\pi^2/3\lambda^2) <r_g^2> \sin^2(\theta/2)] + 2A_2 c \qquad [3]$$

where $<r_g^2>$ is the mean square radius of gyration, independent of molecular conformation, can be used to determine the shape of a molecule if its molecular weight is known. The known or measurable parameters are R_θ, c, K^*, λ and θ. The parameters M_w, $<r_g^2>$, and A_2 can be determined by the method of Zimm (3), which consists of plotting K^*c/R_θ vs $[\sin^2(\theta/2)+kc]$, where k is an arbitrary constant. Extrapolating the data at each concentration to zero angle one obtains a line, the slope of which is the value of A_2. Extrapolating the data at each angle to zero concentration, the slope of the line is a measure of $<r_g^2>$. The intercept of $\theta=0$ and $c=0$ is the value of $1/M_w$ (Figure 1). When using size exclusion chromatography with in-line light scattering detection the inverse of equation [3] is used to calculate the molecular weight of each peak slice:

$$(R_\theta)/(K^*c) = M_w[1 - (16\pi^2/3\lambda^2) <r_g^2> sin^2(\theta/2) - 2A_2cM_w] \qquad [4]$$

The plot of R_θ/K^*c vs. $sin^2(\theta/2)$ at low angles yields a straight line, the intercept of which is weight average molecular weight of that slice.

Figure 2 shows the instrumentation setup. The analytical system consists of an HPLC pump and autosampler (Hewlett Packard Series 1050) with 0.2µm and 0.1µm filters in tandem installed between the two modules, a UV/VIS absorption detector (HP1050 Series VW detector, 79853A), a multi-angle laser light scattering photometer (DAWN DSP-F, or miniDAWN, Wyatt Technology Corporation), and a refractive index detector (HP 1047A DRI detector). A single PC computer is used to control all the modules and for data acquisition and reduction. When there is no chromatographic separation required the chromatographic system is replaced with a

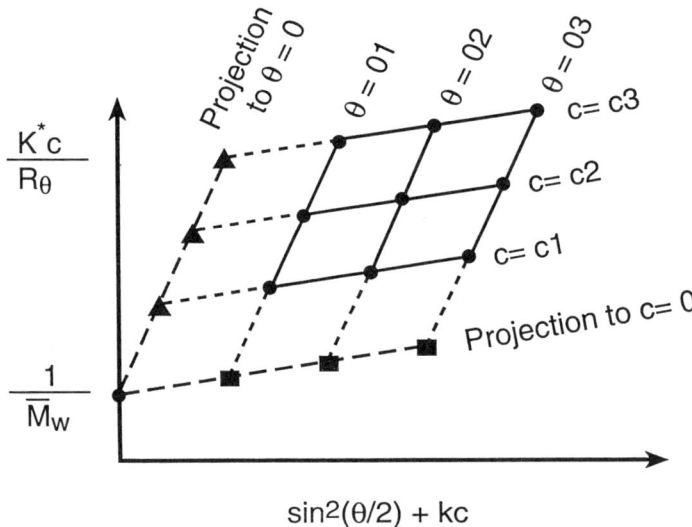

Figure 1. Zimm Plot for Extrapolation of Light Scattering Data. Weight average M_w is obtained by extrapolation to $c = 0$, and $\theta = 0$.

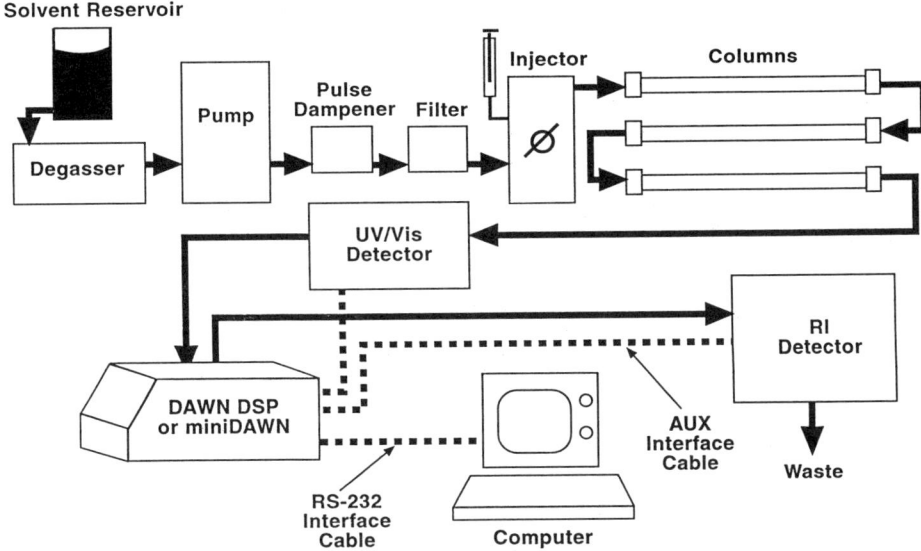

Figure 2. Instrumental Setup for Molecular Weight Determination by Laser Light Scattering Photometry.

manual sample injector and a syringe pump to supply buffer (batch mode). However, in most cases isocratic size exclusion chromatography is used to improve the homogeneity of samples being analyzed (4, 5). This technique can also be used with reverse phase and ion exchange chromatography, although the mobile phase gradient has to be limited to ensure an acceptable *dn/dc* shift throughout the course of the analysis. By having both UV and DRI detectors in-line additional information on protein-nonprotein conjugates can be obtained.

A typical graphic presentation of the experimental results is shown in Figure 3. The upper panel is the chromatographic profile with all three detectors (the delay volumes between detectors were corrected, and the signals were normalized for the main peak): A = light scattering intensity at 90^0, B = RI signal, C = absorption at 230nm. The ratios of the light scattering intensity to Abs_{230} and to RI of the first two peaks are higher than that of the main peak. Since the light scattering intensity is proportional to the product of molecular weight and concentration, these data indicate the presence of higher molecular weight species. The lower panel shows the calculated weight average molecular weight of the protein in each peak. In this example there are at least three different molecular weight species. SDS-PAGE of the sample shows a single band with molecular weight of approximately 18 kD suggesting the presence of monomers and non-covalently associated dimers, and tetramers in the solution.

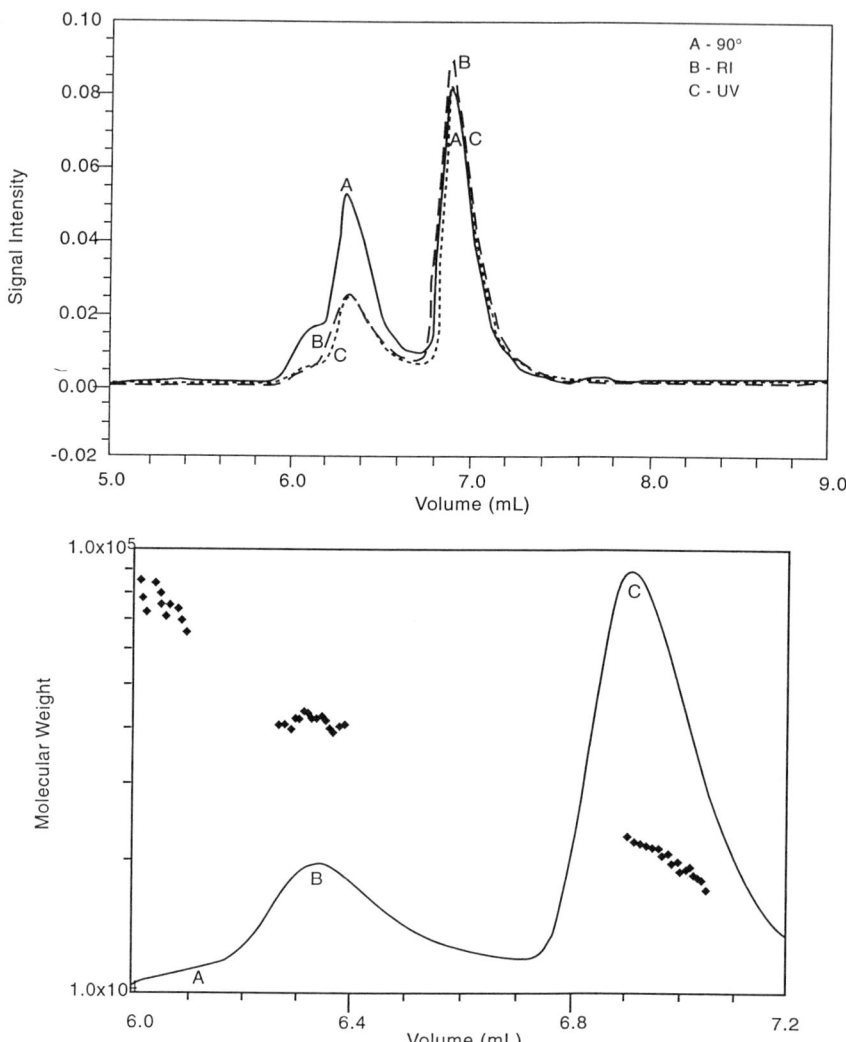

Figure 3. Graphic Presentation of Results from Laser Light Scattering Experiment.
Upper panel: Superimposition of the signals from 90° light scattering (A), RI (B), and UV (C) detectors. Signals were normalized for the main peak.
Lower panel: M_w vs. elution volume. The majority of molecules in peak A, B, and C are tetramer, dimer, and monomer respectively.

Applications

a). Self-associating systems

An important solution property of proteins, especially when comparing analogues, is concentration-dependent self-association. In Figure 4a, protein amounts ranging from 7.5 to 120 µg (0.1 to 2.4 mg/mL) were injected onto a 7.8 mm x 30 cm TSK GEL G3000SW$_{XL}$ column (ToSoHaas). Protein recovery from the column is close to 100%, there is very little shift in elution position, and there is little indication of association based on molecular weight distribution. On the other hand, when an analogue of this protein differing by ten amino acids was analyzed under the same conditions it behaved quite differently. The results are presented in Fig. 4b. Molecular self-association is evident even at very low protein concentrations. The interactions are rapid and reversible.

b). Selection of optimal solvent conditions

Batch mode analysis can be quite useful when screening for optimal formulations. Figure 5 illustrates the M_w of proteins in solution as a function of pH and solute concentration. A citrate-MES-Tris·HCl (20mM each) buffer system was used to cover the pH range of 3 to 9. The monomer molecular weight of this protein is approximately 16 kD. At pH=3 and pH > 7.5, even at a concentration as high as 1.4 mg/mL, most of the protein is in monomeric form. Between pH 3 and pH 7.5 the protein exhibits a pH and concentration dependent self-association. By changing the buffer ions or by adding excipients, the degree of protein association is modified as evidenced in the data curve labeled "FB pH 4", which is a pH4, 10mM NaOAc buffer containing 5% sorbitol. During the development of a protein formulation it is important to know the aggregation characteristics of the protein. This includes factors such as concentration dependence, aggregate solubility, and reversibility of the complex. Figure 6 shows the results of a study of concentration-dependent self-association. Different preparations of a pharmaceutically active protein were formulated at 1.5, 20, and 50 mg/mL in the same buffer system. Aliquots from each solution were diluted to several different concentrations and the M_w of each were determined within 90 minutes of dilution. The dissociation of the protein aggregates was incomplete even at 0.15 mg/mL. When the diluted solutions were allowed to stand at 4°C overnight dissociation was close to complete for those with protein concentration ≤ 1 mg/mL.

c). Kinetics of molecular interactions

The solution behavior of this protein was further studied by coupling size exclusion chromatography with light scattering in order to identify the molecular species involved and to learn how they interact with one another. Figure 7 summarizes the results of six chromatographic separations of 50µg injections of

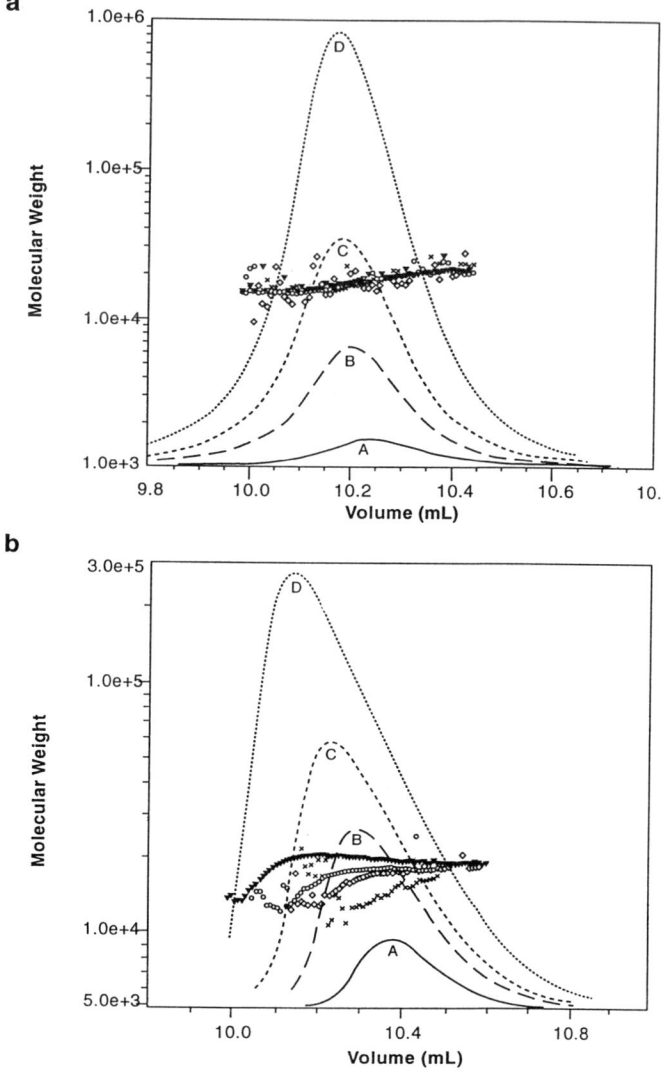

Figure 4. Elution Profile And Molecular Weight Distribution as A Function of The Amount of Protein Injected.
a). A recombinant protein: The amounts(grams) of protein injected are: A: 7.5×10^{-6}; B: 3.0×10^{-5}; C: 6.0×10^{-5}; and D: 1.2×10^{-4}.
b). An analogue of the protein in a): The amounts(grams) of protein injected are: A: 1.2×10^{-5}; B: 3.6×10^{-5}; C: 6.0×10^{-5}; and D: 1.2×10^{-4}.

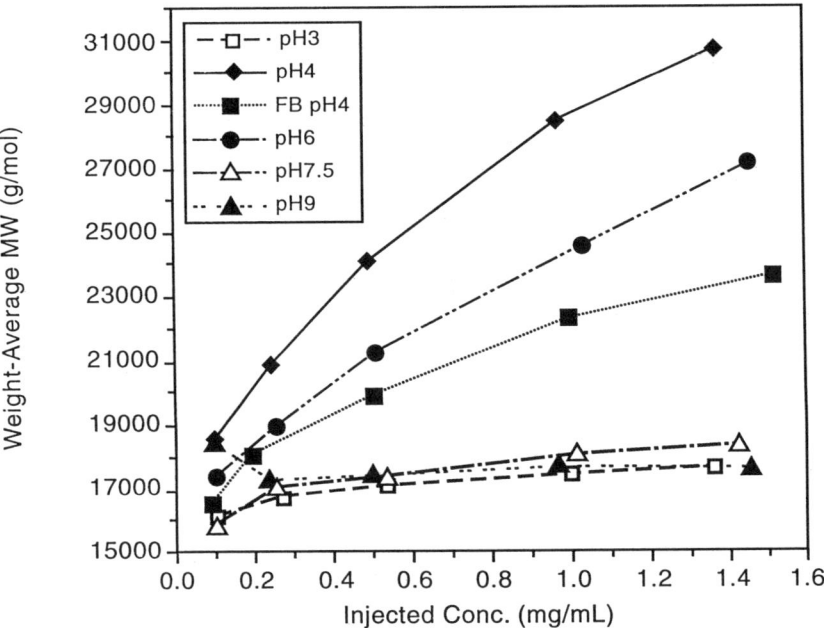

Figure 5. Molecular Weight vs. Concentration as A Function of pH.

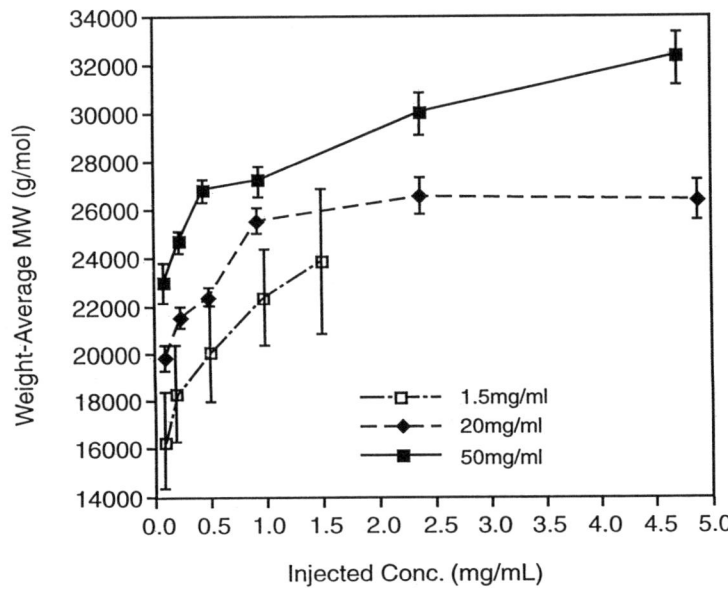

Figure 6. Weight Average Molecular Weight vs. Concentration of The Protein Solution Injected.

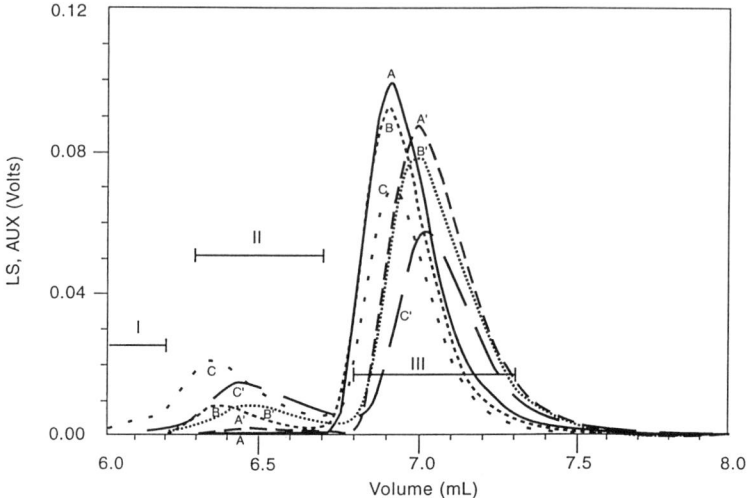

Figure 7. Chromatograms And Molecular Weight Distribution Profiles of A Protein Formulated in Three Different Concentrations. 5, 20, and 50 mg/mL for A, B, and C; For A', B', and C', the protein concentration was 1 mg/mL, each diluted from solution for A, B, and C respectively. Same injection amount for all. Three groupings of association state were assigned as I: "tetramer"; II: "dimer"; and III: "monomer".

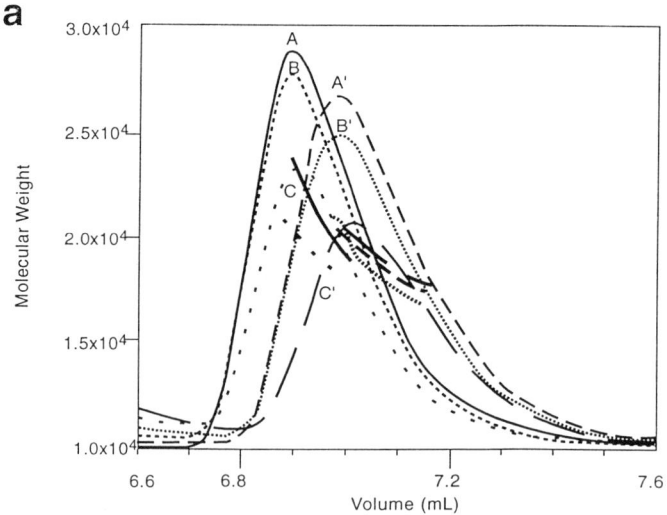

7a. The Main Peak in Figure 7 Expanded to Show The Molecular Association.

protein at varying concentrations. The solid lines are the 90° light scattering intensity traces and the various symbols are the calculated M_w of the protein at each point. The protein concentrations for chromatograms A, B, and C were 5, 20, and 50 mg/mL respectively. For chromatograms A', B', and C' the protein solutions were each diluted to 0.5 mg/mL and held at 4°C for less than 90 minutes prior to the injection. Assuming 0.19 mL/g as the dn/dc value for this protein, the recoveries of the undiluted solutions were approximately 90%. From the chromatograms it appears that the protein population can be divided into three groups eluting at 5.9-6.2 mL(peak I), 6.3-6.7 mL(peak II), and 6.8-7.3 mL(peak III). The weight average molecular weight of each group is slightly higher than that calculated for tetramer, dimer, and monomer respectively, with most of the protein in the last group. The separation into three distinct groups indicates that there is relatively slow or no conversion among the groups during the chromatography. The slope of the molecular weight distribution (scale expanded to show more clearly in Figure 7a), on the other hand, indicates reversible association within each group. (In Figure 7, the relatively sharp upward trend of the M_w distribution curve at the front shoulder of the main peak was caused by dilution and mixing of solution as it left the light-scattering detector and entered the RI detector). Upon dilution to 0.5 mg/mL the elution volume of each peak increased, the slopes of the molecular weight distribution in the main peaks decreased, and the areas of the higher molecular weight peaks decreased with a concomitant increase in the main peak area. The changes are most pronounced for the solution formulated at 50 mg/mL. Even after dilution to 0.5 mg/mL those samples originating from the 20 and 50 mg/mL solutions still contained more of the dimer-like and tetramer-like groups than the sample derived from the 5 mg/mL formulation.

In order to study the kinetics of the inter-species conversions a 50 mg/mL solution of protein was diluted to 1.0 and 0.25 mg/mL. Both solutions were kept on ice until being chromatographed at room temperature. Samples were taken at intervals over a 24 hours period. For the 1.0 mg/mL solution, 50 µg was the injection amount, and for the 0.25 mg/mL solution, 25 µg. The results of this time course experiment are presented in Figure 8. For the 1.0 mg/mL data (Figure 8a), at time T_0 there were three different weight average molecular weight groups of protein molecules (each slightly larger than that of tetramer, dimer, and monomer respectively as in Figure 7) that had been separated by the column. At 3.4 hours post-dilution there were changes in the relative sizes and the retention times for all three peaks (T_1 trace). Both the "tetramer" and "dimer" peaks diminished in areas, however, the "monomer" peak area increased and eluted earlier when compared with the corresponding peak in the T_0 trace. Apparently, as a consequence of dilution, some of the protein in the "tetramer" peak dissociated into dimers and, similarly, those in the "dimer" peak further dissociated into monomers. The main peak must contain species of molecules that were on their way to become monomers. However the kinetics appeared to be slow enough that the non-monomeric species were discernible by SEC. Diluted sample after 8 hours

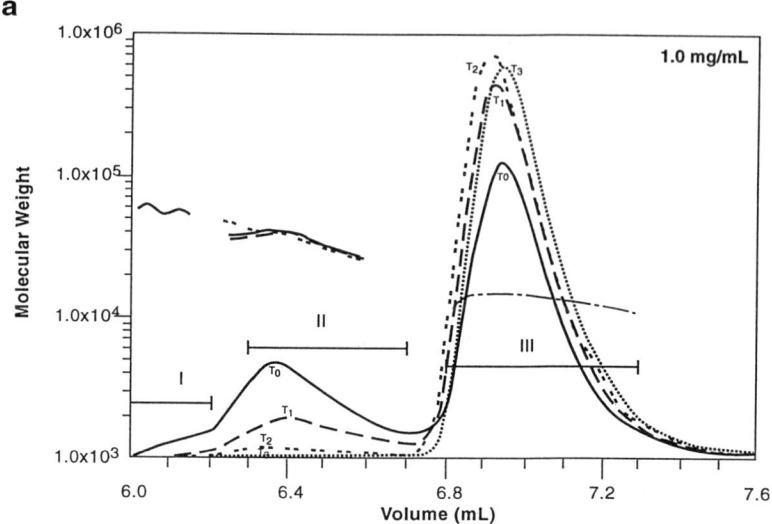

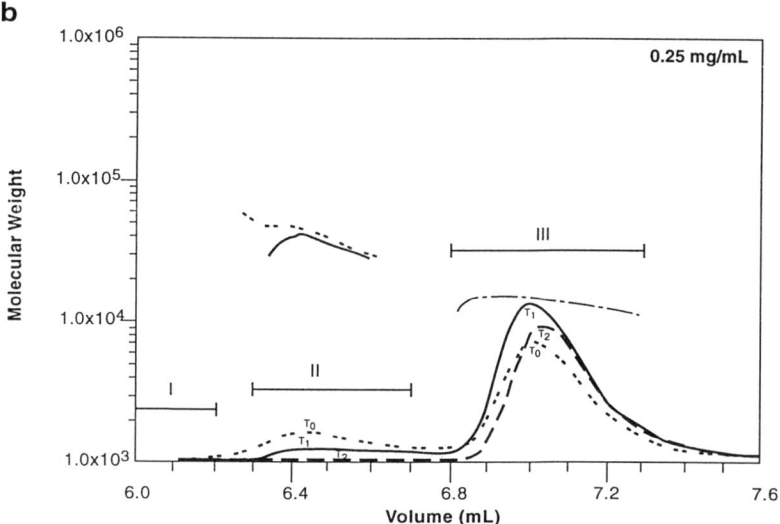

Figure 8. Kinetics of The Dissociation of Protein as A Function of Concentration. Starting protein concentration was 50 mg/mL.
a). Diluted to 1 mg/mL. T_0=0 h; T_1=3.4 h; T_2=8 h; T_3=23.7 h. on Ice.
b). Diluted to 0.25 mg/mL. T_0=0 h; T_1=2.9 h; T_2=8 h. on Ice.
Peak identifications: I: "tetramer"; II: "dimer"; III: "monomer".

incubation in ice showed a similar SEC pattern (T_2 trace) but it appeared that most of the "tetramer" and "dimer" material had gone to the "monomer" peak which eluted even earlier than those in T_1. After incubation on ice for 24 hours after dilution, almost all protein molecules were dissociated into "monomer" as shown in trace labeled T_3. Not only did the the M_w of the main peak approach that of the expected monomer but the elution volume was increased and there were only trace amount of "dimer" present. It took 9 hours at 5° for the 0.25 mg/mL solution to reach a similar state. The results were shown in Figure 8b. For a solution diluted to 1 mg/mL and incubated at room temperature, approximately 8 hours was needed to complete the dissociation process. The information obtained from this provided a guideline for sample preparation in both production and research.

d). Characterization of covalent conjugates in pharmaceutically active proteins

Light scattering can also be a useful tool for characterization of covalent conjugates of protein and methoxypoly(ethylene glycol) (mPEG). Reductive amination of PEG-aldehyde with lysine amino groups can be used as a direct route for conjugate formation (6). However the specificity and degree of PEGylation depend very much on the reaction conditions as well as the properties of the protein. In order to study the PEGylation and to use the PEGylated protein as a pharmaceutical the purified reaction products have to be characterized. The physico-chemical properties of the two components are quite different. Not only there are differences in the absorption spectra and *dn/dc* but more importantly, the conformation of mPEG in aqueous solution, most likely a partial random coil with numerous helical sections (7), is very different from a compact globular protein. The results of SEC of 12 kD mPEG (Figure 9, peak A), Granulocyte Colony Stimulating Factor (GCSF, 18.8 kD, peak B), and their conjugate (monoPEG-GCSF, 30.8 kD, peak C) are shown in Figure 9. Notice the order of the peak elution: 12 kD mPEG was eluted first, even earlier than that of the covalent protein-mPEG conjugate. Both RI and Abs_{280} were used in order to derive the stoichiometry of the protein-mPEG conjugate. This is possible since mPEG has no absorbance at 280nm but has a *dn/dc* value quite different from that of protein.

Two methods were used to determine the molecular weight of the conjugate and its composition. In the first method the values of *dn/dc* for the protein and the mPEG were determined separately. In cases where the isolation of a pure, single species of molecule was not possible it was necessary to assume that the *dn/dc* value of the conjugate is additive. Since the molecular weights and *dn/dc*'s of the protein and mPEG are known, the *dn/dc* of the conjugate was calculated based on an assumed molar fraction of each component. This number was used for molecular weight determination from the light scattering intensities of the conjugate peak fractions. The iteration stopped when the calculated and the expected molecular weight matched. Table I lists the values of the calculated *dn/dc* and the expected and determined weight average molecular weights of mono-PEGylated GCSF's produced from different molecular weight mPEG's. The discrepancies between the

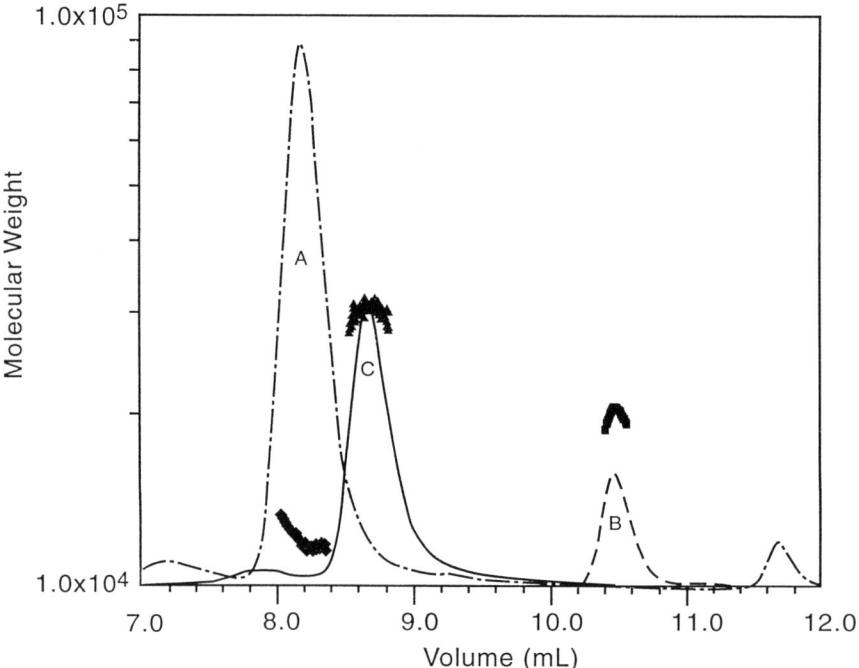

Figure 9. Size Exclusion Chromatography of mPEG (A), GCSF(B), And mono-PEG-GCSF (C) on A TSK GEL G3000SW$_{xl}$ Column.

predicted and the observed values are within the range of the heterogeneity of the mPEG starting samples. Other techniques such as gel electrophoresis and laser desorption mass spectrometry verified the light-scattering results.

The second experimental approach depends on a chromatographic separation with well resolved peaks. The amount of protein in the conjugate is determined from its Abs_{280} area in the chromatogram and then used to calculate the protein contribution to the RI peak of the conjugate. Subtraction of this amount from the actual RI peak area will give the RI contribution from the mPEG. This value, in turn, is used to calculate the weight amount of mPEG. By converting both weight amounts into molar quantities one can determine the molar ratio of mPEG to protein, thus the degree of PEGylation of the conjugate (7). This method requires quantitation of protein and mPEG through constructing standard curves of Abs_{280} and/or RI signals for each component. This is very useful where multiple molecular species are present in a sample being analyzed and the determination of dn/dc for each species is impossible. Examples of data analysis using this method are shown in Figure 10. Once the molar ratio of mPEG to protein is obtained the molecular weight is determined by method one described above to confirm the result. Application of this method showed that the deviation from the expected molecular weight of the various conjugates was small.

This method can be applied to a number of systems. As long as one component of a conjugate, regardless of its chemical nature, is quantifiable, and the presence of the other does not interfere with the quantification of that component in the conjugate, the stoichiometry is obtainable in principle.

e). Characterization of protein-ligand interactions

The specificity of a protein-ligand interaction depends on the chemical integrity and maintenance of the active physical state of the protein molecules in solution. For example, the active forms of nerve growth factor, brain-derived neurotrophic factor, and stem cell factor are all non-covalent dimers. The molecular weight determination will confirm the N-mer state of the protein under the conditions studied. The interacting system shown in Figures 11a though 11c illustrates the binding of a soluble receptor to a protein in phosphate buffered saline (PBS). In each figure the upper panel shows the 90° detector's light scattering intensity trace while the lower panel shows the Abs_{280nm} and RI traces in the voltage responses. Figure 11a is a chromatogram of the protein alone; Figure 11b is a chromatogram of the soluble receptor by itself. The chromatogram in Figure 11c was generated from the incubation mixture of the soluble receptor with molar excess of protein. Based on the calculated molecular weights, the protein was found to exist as a trimer in PBS while the receptor is monomeric, and each receptor binds to a protein trimer (see Table II). Without the light scattering intensity data it is necessary to run a series of experiments varying the protein:receptor ratio in order to determine the stoichiometry of the complex. However, with incorporation of the

Mw.PEG (kD)	Wt. PEG/Wt.GCSF	[PEG]/[GCSF] (M/M)
6	0.3186	0.9982
12	0.7049	1.1044
20	1.1325	1.0645
25	1.2755	0.9592
12	0.6404	1.0032
12 #	1.2642	1.9805
12 ##	0.1227	0.1923

Peak A in the early fraction; ## Peak C in the late fraction, of ion-exchange chromatography as shown below. M_W calculations from ASTRA, using calculated dn/dc values for individual molecules, confirmed the degree of pegylation in each case except Peak C material, which had been identified (by de-pegylation and mass spectrometry) to be truncated PEG (Mw around 3kD) bound to GCSF.

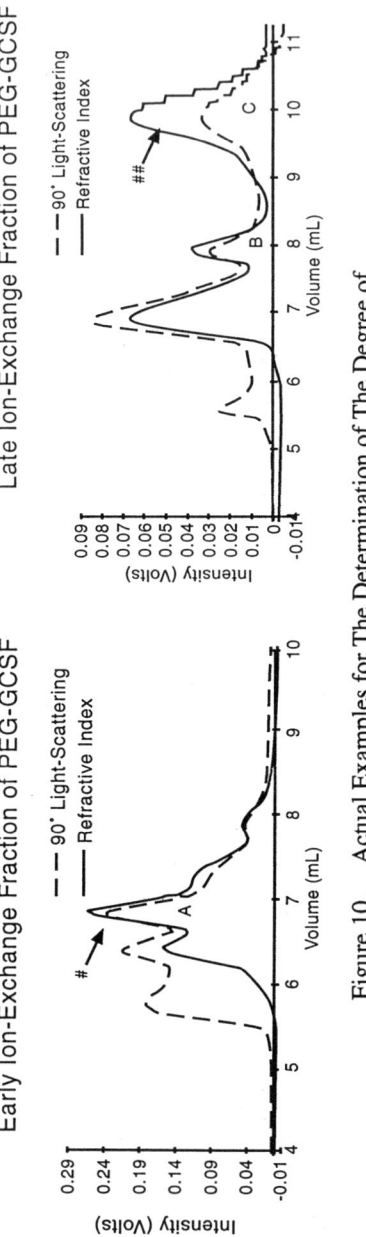

Figure 10. Actual Examples for The Determination of The Degree of PEGylation for Partially Purified PEGylated Products Using Area Integration Method.

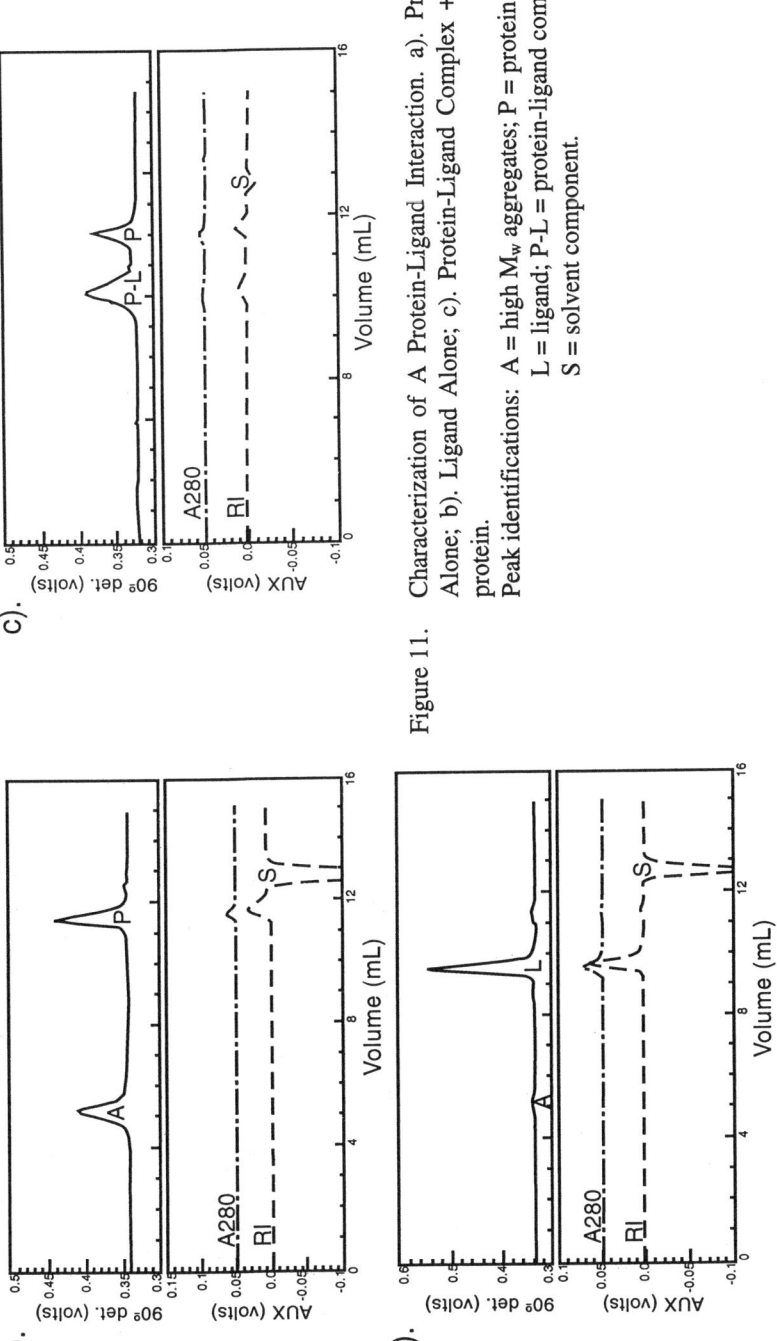

Figure 11. Characterization of A Protein-Ligand Interaction. a). Protein Alone; b). Ligand Alone; c). Protein-Ligand Complex + free protein.
Peak identifications: A = high M_w aggregates; P = protein; L = ligand; P-L = protein-ligand complex; S = solvent component.

Table I. Molecular Weights of Purified mono-PEG-GCSF.

Mw, PEG (kD)	Mw, PEG-GCSF, expected (kD)	dn/dc, calc. (mL/g)	Mw, PEG-GCSF, obtained (kD)
6	24.8	0.178	25.3
12	30.8	0.168	31.4
20	38.8	0.161	39.2
25	43.8	0.157	43.3

Table II. Molecular Weights of The Interacting Species in Figure 11.

Panel	Peak	Mw (kD) determined	Mw (kD) of monomer	Molecular species
11-a	P	52.84	17.35	P_3
11-b	L	60.32	56.00	L
11-c	P-L	115.90	?	P_3L
11-c	P	52.77	17.35	P_3

light scattering detector, the molecular weight and therefore the stoichiometry of the complex can be determined directly. In this example both components were proteins and all three species were separable by HPLC thus making the application simple and straightforward. Studies of nucleic acid-protein, antibody-antigen, dye-protein and vesicle-protein interactions are some additional applications.

Summary

The ability to determine the solution structure of pharmaceutically active proteins is very important in the biotechnology industry. When a multi-angle laser light scattering detector is used in series with UV/VIS and differential refractive index detectors absolute molecular weight, radius of gyration, and other structural features can be readily determined. Because of the relatively small sizes and low concentrations of proteins in our studies, the applications presented are based mainly on molecular weight information rather than root mean square radius. Potential applications include:

- Detection of the presence of irreversible, soluble aggregates.
- Identification of reversible self-associating system including description of the kinetics and the factors affecting the reaction.
- Characterization of the covalent conjugation products of a protein with either protein or non-protein species.
- Determination of the stoichiometry of protein-ligand interactions.

Literature Cited

1. Wu, H. *Chinese J. Physiol.* **1931**, *vol. V, No. 4*, p. 321. [Reprinted in "Protein Stability," *Advances in Protein Chemistry* **1995**, *vol. 46*, p.1]
2. Tanford, C. *Physical Chemistry of Macromolecules*; John Wiley & Sons, Inc.: New York, 1961; pp 275-316.
3. Zimm, B.H. *J.Chem. Phys.* **1948**, *vol. 16*, no. 12, pp 1093-1099. [Reprinted in *Light Scattering from Dilute Polymer Solutions,* McIntyre, D; Gornick, F, Eds.;Gordon and Breach, New York, NY, 1964, pp.149-156.]
4. Takagi, T. *J. Biochem.* **1981**, *vol. 89*, pp. 363-368.
5. Maezawa, S.; Takagi, T. *J. Chromat.* **1983**, *vol. 280*, pp.124-130.
6. Harris, J. M.; Macromol, Rev. *Chem. Phys.* **1985**, *vol. 25*, pp. 325-373.
7. Kunitani, M.; Dollinger, G.; Johnson, D.; Kresin, L. *J. Chromat.* **1991**, *vol. 588*, pp. 125-137.

Chapter 11

Applications of Ultraviolet Absorption Spectroscopy to the Analysis of Biopharmaceuticals

Henryk Mach[1], Gautam Sanyal[1], David B. Volkin[1], and C. Russell Middaugh[2]

[1]Vaccine Pharmaceutical Research and [2]Human Genetics, Merck Research Laboratories, WP78-302, West Point, PA 19486

The introduction of computer-controlled diode-array spectrophotometers has permitted fast acquisition of highly reproducible UV spectra of proteins and peptides. Furthermore, immediate data analysis is now feasible with specific computer software. This non-destructive technique, which depends on contributions from UV absorption of aromatic amino acids, cystine and light scattering, can be used for concentration determination and detection of impurities as well as assessment of aggregation state. The use of second derivative near-UV spectra also permits sensitive detection of proteins in nucleic acid samples. The high reproducibility of the second derivative band positions and their dependence on chromophore microenvironment also provides a fast and sensitive probe of protein structural integrity.

Four decades ago, UV spectroscopy, despite being technically less than optimal, offered one of the few opportunities to study the environment of aromatic amino acids, and indirectly, protein structural transitions (1). More powerful methods to analyze protein structure have emerged since then such as X-ray crystallography, nuclear magnetic resonance (NMR), circular dichroism (CD), fluorescence, Fourier-transform infrared spectroscopy (FTIR) and differential scanning calorimetry (DSC). Importantly, however, UV spectrophotometers are still in use in virtually every biochemistry laboratory, because of their simple, rapid and nondestructive capability to quantitate protein and nucleic acid concentrations. The development of fast and highly precise diode-array spectrophotometers as well as computers that employ powerful algorithms to extract the information contained in UV spectra has resulted in renewed interest in this technique as a tool in the evaluation of protein structure and stability. In addition to concentration determination, UV spectra can provide quantitative information about proteins such as the content of aromatic amino acids (2-4) and degree of solvent exposure (5,6), as well as thermodynamic parameters of conformational transitions (7). In addition, shifts of aromatic amino acid bands may reveal the polarity of side chain aromatic microenvironments and structural changes induced by the presence of various binding entities. Furthermore, the light scattering component of UV spectra can be useful in assessing the aggregation state of macromolecules (8). Since unfolding triggers the aggregation of many proteins, the kinetics of turbidity appearance is a sensitive probe that can be used for the screening of stabilizing agents in pharmaceutical applications (9). Thus UV spectrophotometry offers a rapid, accurate and inexpensive alternative to less readily available high-resolution spectroscopic techniques.

UV Absorbing Chromophores in Proteins and Nucleic Acids

The aromatic amino acids, tryptophan, tyrosine and phenylalanine have distinct spectra between 250 and 300 nm. The spectrum of tryptophan in aqueous buffer is shown in Fig. 1 A. It manifests an absorption maximum near 280 nm, with a molar extinction coefficient of 5550 M^{-1} cm^{-1} (10). The incorporation of an aromatic amino acid side chain into the interior of a protein generally results in a decrease in the polarity of its microenvironment. This perturbs the energy necessary for electrons to move to excited states and shifts the position of such absorption bands to higher wavelengths. Although the extent of such shifts is different for each particular side chain in a protein due to the unique charge distribution and spatial arrangement of neighboring residues, it is possible to jointly analyze all of the residues of one type in a group of proteins. Matrix multicomponent analysis, performed on a set of spectra of a number of globular proteins allows extraction of such data. The resulting tryptophan spectral component, which may be thought of as that of indole moieties in a "typical" protein, is shown as a dashed line in Fig. 1 A. Both spectra show the characteristic fine structure arising from discrete electronic transitions in the chromophore. These features can be amplified numerically by calculation of derivative spectra, utilizing general formulas developed for any order which incorporate coefficients specific for order, degree of fitting polynomial and window size (11,12). For near-UV spectra two preceding and two following data points, multiplied by predefined coefficients, added and divided by a predefined factor, are most commonly used (13). The second derivative spectra of both the actual tryptophan spectrum in aqueous buffer and the tryptophan spectrum deconvoluted from the set of 12 spectra of globular proteins (7) are shown in Fig. 1 B. The similarity of the second derivative spectral features validates use of the deconvolution of the zero-order spectra. The increased magnitudes of the second derivative peaks of the of tryptophan residues in the "average protein" reflect the more rigid microenvironment in protein interiors, which results in band sharpening.

The spectra of tyrosine in aqueous buffer as well as in the "average protein" interior are shown in Fig. 2 A. The absorption maximum occurs at about 275 nm for tyrosine in aqueous buffer and at about 277-278 nm in proteins (6). In contrast to tryptophan, marked dependency of the extinction coefficient on the polarity of the microenvironment is apparent. The large shift upon incorporation into protein interiors reveals the high sensitivity of this residue to the polarity of its microenvironment. The lack of band sharpening presumably reflects the greater heterogeneity of the average tyrosine population in proteins. Indeed, due to the presence of both an apolar aromatic ring as well as a polar hydroxyl group, roughly equal numbers of tyrosine residues are often found buried inside proteins and located on their surfaces. In fact, the second derivative spectra actually have lower band magnitudes in the "average protein" than in tyrosine side chains fully exposed to the aqueous buffer (Fig. 2 B).

The spectrum of phenylalanine, although of weak intensity ($\varepsilon=197$ $M^{-1}cm^{-1}$ at 258 nm) (14), possesses well defined fine structure, which results in a very distinct second derivative spectrum (Fig 3 A and B). This residue appears to be less sensitive to the polarity of its environment than either tryptophan or tyrosine. Cystine (disulfide bridged cysteine) residues also posses weak absorbance ($\varepsilon = 134$ $M^{-1}cm^{-1}$ at 280 nm) (15). However, several disulfide bridges in one protein may typically account for a few percent of the total absorbance at 280 nm and thus warrant inclusion in near-UV spectral analyses. Such consideration apply most often to eukaryotic intracellular proteins which generally exist in an oxidizing environment.

Nucleic acids are encountered in pharmaceutical preparations not only as essential components of a formulation (i.e. as components of viruses or as gene therapy or antisense agents) but also as persistent contaminants. These macromolecules possess intense absorbance arising from purine and pyrimidine bases with a maximum near 260 nm but poorly defined (i.e. broad) spectral features (Fig. 4 A). Nucleic acid second derivative spectra, although less intense than aromatic amino acids, have distinct and highly reproducible features (Fig. 4 B). For example, double stranded DNA can be recognized by a series of negative peaks between 250 and 290 nm, while spectra of transfer RNA manifest a single major peak at 258 nm with secondary peaks at approximately 265 nm (16). These features can aid in the identification of unknown nucleic acid preparations, or samples contaminated thereby.

The spectra of aromatic amino acids as well as nucleic acid bases are sensitive to the microenvironment in which these residues are embedded. By fitting of gradually shifted standard spectra to the spectra of a number of proteins it was found that the shift between the spectrum of an aromatic amino acid in water and that of the corresponding residue embedded in a highly apolar protein interior is 5 nm for tyrosine, 4 nm for tryptophan and 2 nm for phenylalanine (6). It was also shown that these spectral shifts depend not only on the degree of solvent exposure, but also on the polarity of adjacent residues. Similarly, purine and pyrimidine bases show large increases in their molar absorptivities upon transition from double-stranded to single-stranded (unfolded) conformations (17), which serves as a basis for analyzing polynucleotide "melting" phenomena.

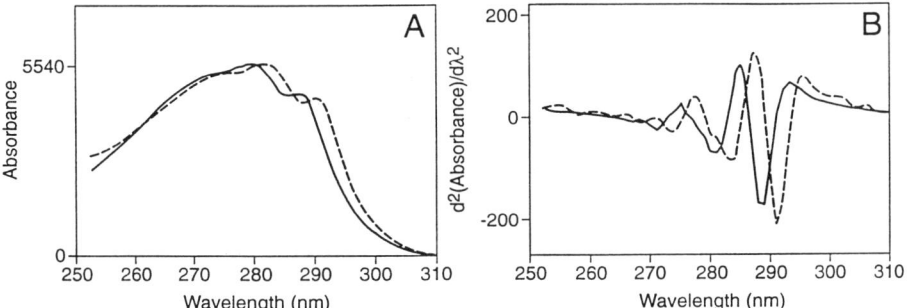

Figure 1.

Normal (zero-order) (A) and second derivative (B) spectra of N-acetyl-L-tryptophanamide (solid line) and that of tryptophan in an "average protein" (dashed line) as determined by multicomponent analysis of a set of 12 globular proteins. The absorbance values are normalized to a 1 molar solution in a 1-cm pathlength cell.

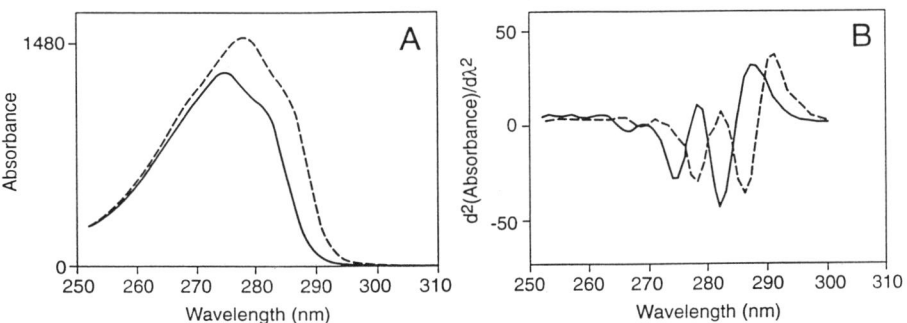

Figure 2.

Normal (zero-order) (A) and second derivative (B) spectra of N-acetyl-L-tyrosinamide (solid line) and tyrosine in an "average protein" (dashed line) as determined by multicomponent analysis of a set of 12 globular proteins. The absorbance values are normalized to a 1 molar solution in a 1-cm pathlength cell.

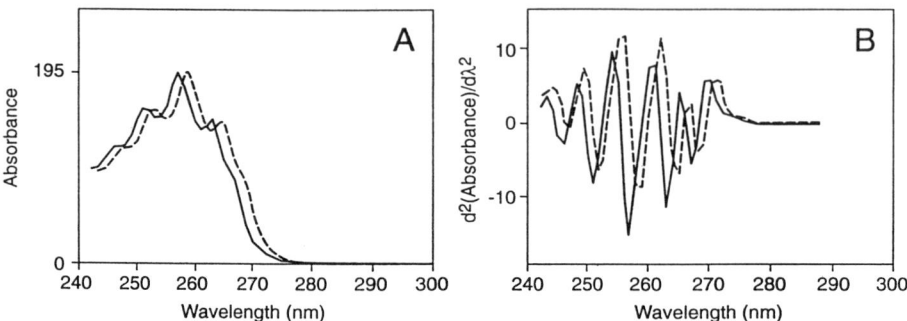

Figure 3.

Normal (zero-order) (A) and second derivative (B) spectra of N-acetyl-L-phenylalanine ethyl ester (solid line) and phenylalanine in an "average protein" (dashed line) as determined by multicomponent analysis of a set of 12 globular proteins. The absorbance values are normalized to a 1 molar solution in a 1-cm pathlength cell.

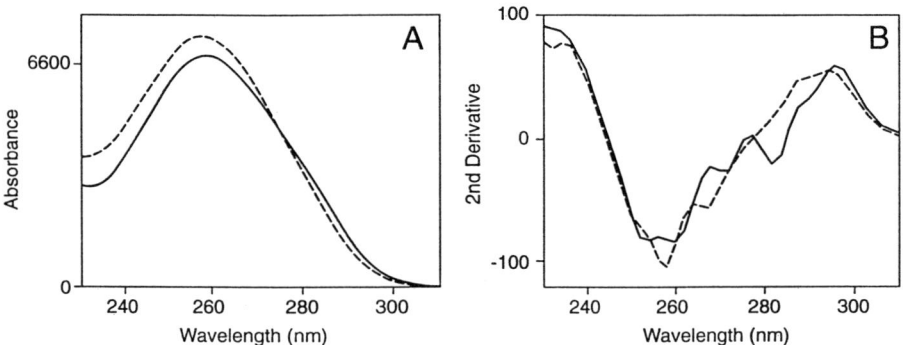

Figure 4.

Normal (zero-order) (A) and second derivative (B) spectra of calf thymus DNA (solid line) and bakers yeast t-RNA (dashed line). The absorbance values are normalized to a 1 molar solution in a 1-cm pathlength cell.

Calculation of Molar Absorptivity From the Protein Sequence

The calculation of extinction coefficients from sequence data was attempted when protein UV spectroscopy was first introduced (1). However, only extinction coefficients of aromatic amino acids in aqueous buffer were available. In addition, chemical methods for the measurement of extinction coefficients, such as dry weight determination or the Lowry (18) and Bradford (19) colorimetric techniques had error levels approaching 10%. As the available data base increased, it became apparent that these calculations were only accurate for tryptophan-rich proteins (20). The use of extinction coefficients of model aromatic amino acid compounds in guanidine hydrochloride was found to produce more consistent results (21). Recently, however, application of matrix deconvolution methods has permitted the determination of these values based on empirical values of extinction coefficients obtained by deconvolution analysis of spectra from a large number of proteins (15). While the extinction coefficient of protein-bound tryptophan was found to be 5540 $M^{-1}cm^{-1}$ (see Fig. 1 A), near previously employed values, the extinction coefficient of protein-bound tyrosine at 280 nm was found to be 1480 $M^{-1}cm^{-1}$ compared to 1280 $M^{-1}cm^{-1}$ in aqueous buffer (see Fig. 2 A). Essentially the same values were later found consistent with an even larger protein data base (22). Thus the extinction coefficient of a globular protein can be predicted with a typical error of less than 2% by the formula (15):

$$\varepsilon = 5540 * N_{trp} + 1480 * N_{tyr} + 134 * N_{s-s} \tag{1}$$

where N_{trp}, N_{tyr} and N_{s-s} denote the number of tryptophan, tyrosine and cystine residues, respectively.

Most of the error in the above equation is introduced by the solvent sensitivity of the tyrosine spectra. Due to its high molar absorptivity, however, the Trp component is usually the dominant one. When tryptophan residues are absent, predictions of extinction coefficients may fail if the average tyrosine microenvironment is different from that of an "average protein". This most frequently happens in the case of peptides or small non-compact proteins. Since tryptophan is absent, the spectral band position of the tyrosine component can be directly determined from near-UV spectra by calculating the first derivative and determining the position of the intersect with the wavelength axis (13). As seen in Fig. 2 A, this peak position is directly correlated to the expected molar absorptivity. The following equation can then be used to calculate the extinction coefficient of tyrosine residues specific for a given protein or peptide:

$$\varepsilon = 50.9 * \lambda_{max} - 12{,}650 \tag{2}$$

This equation was derived using the extinction coefficients and λ_{max} values of tyrosine in aqeous buffer and in ribonuclease A.

These methods are particularly useful for newly obtained protein samples, prior to experimental determination of extinction coefficient most commonly by quantitative amino acid analysis.

Light Scattering Contributions to near-UV spectra

A light scattering component in a UV spectrum can arise from the fact that part of the incident beam can be elastically scattered by large solute molecules and thus does not reach the detector. This process does not involve the absorption of photons by aromatic amino acids and nucleic acids as previously described. By convention the term "optical density " (OD) is used for the sum of absorption and light scattering events. Accurate description of light scattering has two-fold importance. First, it is necessary to subtract any light scattering contribution to obtain true absorption spectra to employ calculated protein molar absorptivities in concentration determination. Second, the intensity of the scattering component contains information about the aggregation state of the solute. Scattering becomes significant when the size of a macromolecule becomes comparable to the wavelength of the light used. The general relationship describing such light scattering is (8):

$$d \ln OD/ d \ln \lambda = a \qquad (3)$$

or, after integration

$$\ln OD = a \ln \lambda + b \qquad (4)$$

where OD is the optical density, λ is the wavelength, a is the slope and b is the intersect. The slope a can vary from a value of -4.0 (Rayleigh scattering for molecules smaller than one-tenth of the wavelength of incident light) to +2.2 in a complicated fashion which can be calculated from Mie theory (particles much larger than λ). A more detailed description of the theory is available (8). The linear relationship between log OD and log wavelength allows one to model a scattering component by extrapolation of OD values outside the absorption range of a protein, i.e. > 320 nm. A simple correction of absorption spectra based on this principle can be chosen in the software controlling most modern spectrophotometers. Alternatively, the light scattering correction at selected wavelengths can be calculated using OD values at 320 and 350 nm employing the formula (13):

$$OD (\lambda) = 10^{(m+1) \log OD320 - m \log OD350} \qquad (5)$$

where m= 64.32 -25.67 log λ (m=1.5 for 280 nm and m=2.33 for 260 nm)

An example of a light scattering correction performed by extrapolation is shown in Fig. 5. Subtraction of the light scattering optical density yields in principle the true absorbance spectrum of the protein component.

Light scattering (turbidity) can also be used as a probe of aggregation. For values of a close to -4.0 (Rayleigh scattering) turbidity is directly proportional to particle mass (8). Extensive aggregation, however, often results in size increases to the extent that the values of a approach 2. Under these conditions turbidity becomes inversely proportional to particle mass. Thus in many aggregating systems after an initial quasi-linear increase of optical density, a plateau and then a decline in optical density is observed. The initial increase in turbidity is frequently used to probe the kinetics of protein unfolding when aggregation accompanies structural changes. An example of the use of this approach to screen the ability of a sulfated polysaccharide (heparin) to stabilize a protein drug (acidic fibroblast growth factor) is shown in Fig. 6 (9).

Multicomponent Analysis

The typical analytical approach to multicomponent analysis UV spectra employed prior to the ready availability of personal computers involved the use of a system of simultaneous equations employing equal numbers of spectral data points acquired at wavelengths characteristic of the analyzed components. The use of computer software now allows one to utilize an entire spectrum (an overdetermined system) (23). In addition, since the acquisition of an entire spectrum may require only 0.5 second, the standard deviation of the measurement at each data point (employing diode-array technology) is available. This information can then be incorporated into calculation by weighting the data points according to their relative noise levels (maximum likelihood method). Least-squares analysis of the spectrum of a mixture requires the standard spectra of the individual components. The actual spectra of the aromatic amino acids as they exist in proteins, however, are not explicitly available. These spectra can be estimated by deconvolution of a set of protein spectra of known amino acid composition as described above to obtain the band positions for "average" protein-resident aromatic amino acids residues. It is generally not practical, however, to perform such analysis for every sample. We find it convenient to select a number of common proteins which have aromatic amino acid microenvironment representative of the entire protein population. Such a set is presented in Table I. The concentrations of the constituent chromophores in each calibration spectrum are also given, with lysozyme chosen to provide a tryptophan and ribonuclease a tyrosine standard. N-acetyl-phenylalanine ethyl ester in 100% ethanol was found to be a good approximation for phenylalanine residues embedded in protein interiors. DNA is also included to complement the data set used in multicomponent analysis because of its frequent presence as a significant contaminant. The calibration step can be described by the simple matrix representation:

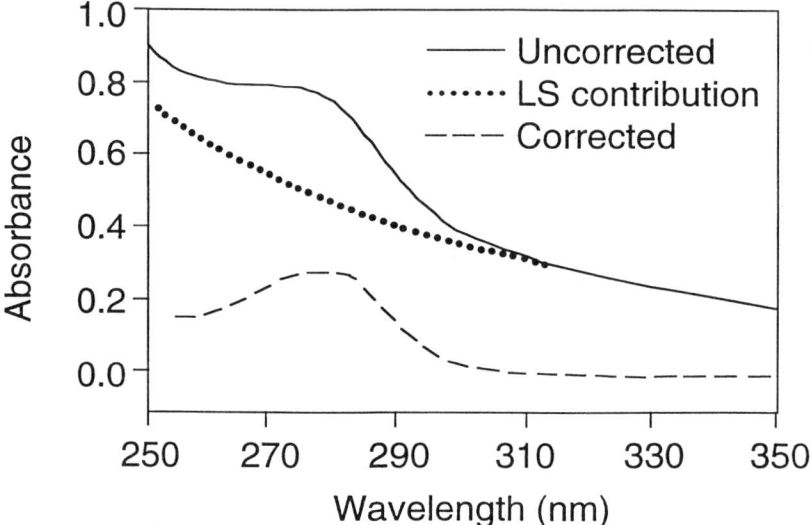

Figure 5.

Near-UV spectrum of a 0.1 mg/ml solution of PRP-OMPC (conjugate of polyribosyl ribitol phosphate and the outer membrane protein complex from Neisseria meningitidis) (solid line). The light scattering component was removed by subtraction of extrapolated data (dotted line) from a linear plot of log OD vs log wavelength between 320 and 350 nm, and the resultant corrected spectrum is shown as dashed line (reproduced with permission of Eaton Publishing Co., 154 E. Central St., Natick, MA 01769-3644 USA from Mach & Middaugh, August 1993, **BioTechniques** 15: 240-242).

Table I. Standard concentration matrix containing concentrations of aromatic amino acids and DNA in multicomponent analysis calibration standards (see text for details).

	Concentrations (mM)			
$A_{280}=1.0$	Trp	Tyr	Phe	DNA
lysozyme	157	78	78	0
RNAse A	0	637	319	0
N-PHE	0	0	5130	0
DNA	0	0	0	152

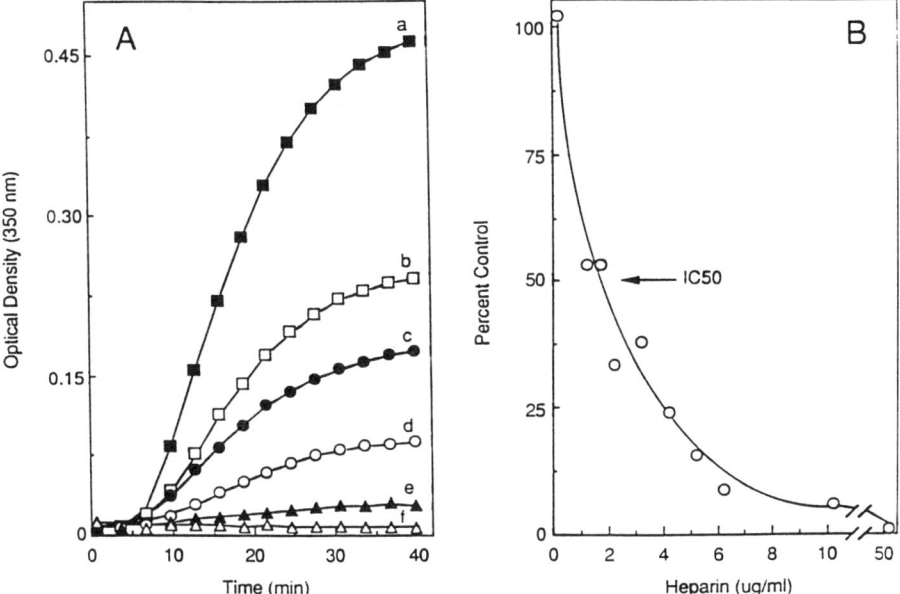

Figure 6.

Effect of heparin concentration on the heat-induced aggregation of acidic FGF at 40 °C. (A) Time course of turbidity formation at varying concentrations of heparin: (B) Effect of heparin on thermal stability at 40 °C. The Y-axis represents the maximum rate of turbidity formation normalized to sample (a) containing no heparin (adapted from ref. 9 with permission).

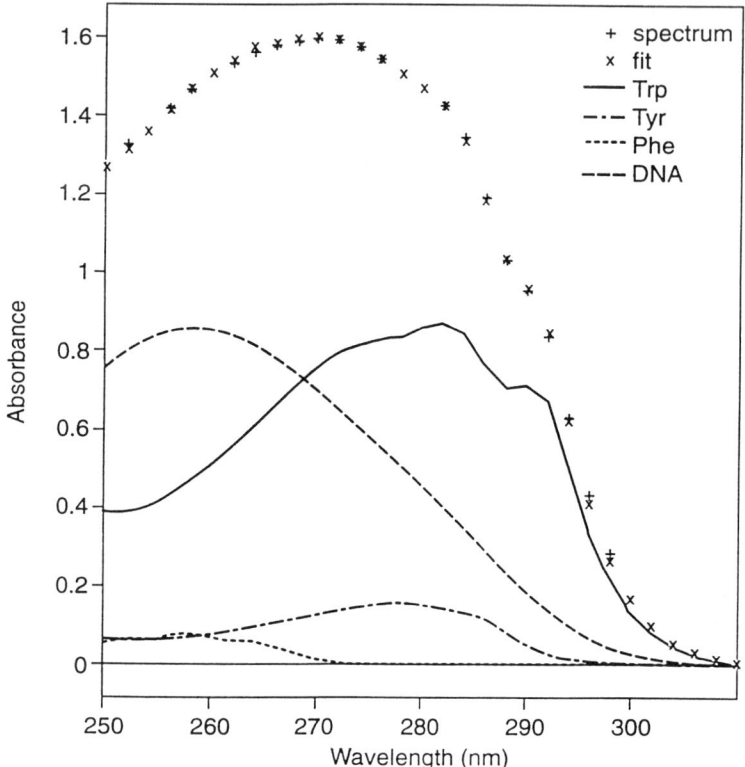

Figure 7.

Multicomponent analysis of the zero-order spectrum (denoted by +) of a mixture of DNA and β-lactoglobulin after light scattering correction. The fitted component spectra are shown: Trp (solid lines), Tyr (dash-dotted lines), Phe (dotted lines) and DNA (dashed lines). The sum of the fitted components is denoted by (x). All spectra were acquired with a Hewlett-Packard 8452A spectrophotometer using 1 cm quartz cell and 0.5 s data acquisition time.

$$F = HC_S \tag{6}$$

where F is the matrix containing the set of calibration spectra (lysozyme, ribonuclease A, N-Phe and DNA), H is the matrix of pure component spectra and C_s is the standard concentration matrix (Table I). The values of matrix H are obtained by rearranging equation (6):

$$H = FC_s^T(C_sC_s^T)^{-1} \tag{7}$$

where T denotes matrix transposition and -1 denotes matrix inversion.

Once the spectra of the pure component are available, they can be used in the matrix expression describing their composition in the sample (unknown) spectrum:

$$S = HC \tag{8}$$

where S is the sample spectrum and C is the concentration of each analyte.

The solution is given by:

$$C = (H^TH)^{-1}H^TS \tag{9}$$

Once the UV spectra of lysozyme, ribonuclease A, N-acetyl-phenylalanine ethyl ester and DNA are measured and stored in the computer memory, the multicomponent analyses described above require only several seconds of computing time and can be completed automatically after each measurement. An example of such an analysis is shown in Fig. 7A. A mixture of β-lactoglobulin and DNA was prepared and its UV spectrum analyzed for aromatic amino acid and DNA content. The experimental spectrum could be accurately reproduced using standard spectra obtained as described above. A light scattering correction was performed prior to the analysis. A multicomponent analysis of the second derivative spectra of the same protein/DNA mixture is shown in Fig.8. In this case a correction for light scattering was not necessary, since the second derivative of the light scattering component is essentially zero, as a consequence of its lack of inflection points (24).

Standard reference UV spectra are useful when the UV spectrum of a pure protein is not known or the protein is not pure, in which case the constituent aromatic amino acids are directly determined. When the spectra of the individual proteins in a mixture are known, multicomponent analysis can be used to

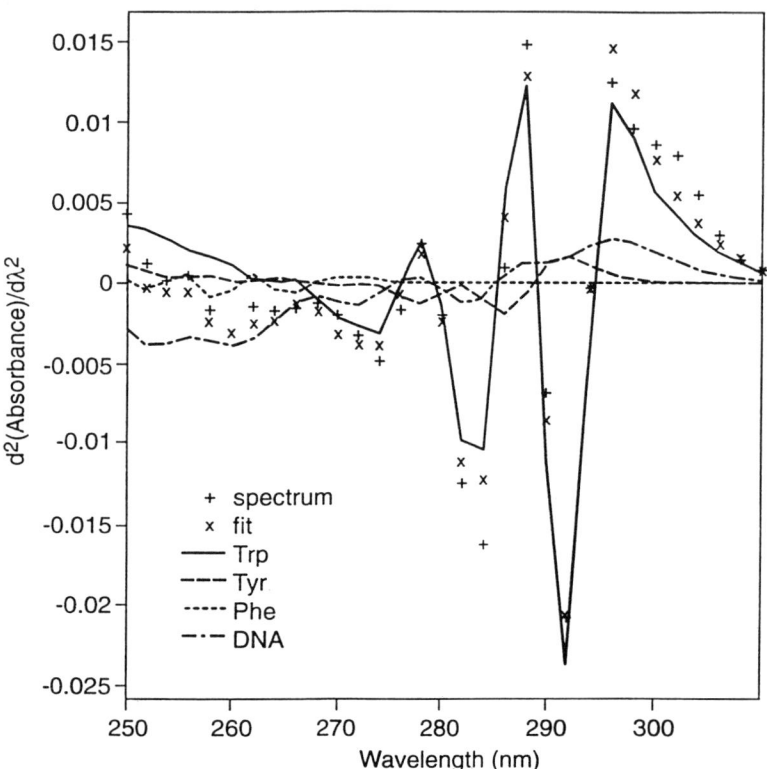

Figure 8.

Multicomponent analysis of the second derivative spectrum (denoted by +) of a mixture of DNA and β-lactoglobulin. The fitted component spectra are shown: Trp (solid lines), Tyr (dashed lines), Phe (dotted lines) and DNA (dash-dotted lines). The sum of fitted components is denoted by (x).

deconvolute the composite spectrum, provided that the aromatic amino acid compositions of each protein are different (25). Since three independent components contribute to the standard UV spectra (viz. Trp, Tyr and Phe), at most three component protein mixtures can be resolved in this manner. The DNA spectrum used in the above example can also be substituted or supplemented by additional components expected to be present in the analyzed solutions. These may include surfactants, stabilizing and antimicrobial agents, etc. The results of multicomponent analysis include not only confidence levels for the concentration result of each analyte, but also a factor describing the independence of standards. These parameters aid in assessing the reliability of the analysis.

Quantitative Analysis of Protein and Nucleic Acids Mixtures

The use of second derivative spectra has the advantage of removing smooth spectral components such as light scattering or baseline drifts and amplifying fine features of electronic or vibrational origin, while maintaining the validity of the Beer-Lambert law of the proportionality between absorbance and concentration. The calculation of the second derivative does not, however, introduce any new information into a spectrum. While it selectively removes unwanted information, it also decreases the signal-to-noise ratio. As an alternative, zero-order spectra can be directly fit to a combination of component spectra. It is not obvious, therefore, which analysis (zero-order or second derivative) is more suitable to analyze mixtures of proteins and nucleic acids, especially when one of the components is present in small amounts.

To clarify this issue we first serially diluted a protein (β-lactoglobulin) into a solution of constant DNA concentration and plotted the measured amounts (as probed by the concentration of the Trp component) against the amount predicted from serial dilution (Fig. 9). A Hewlett-Packard 8452A diode-array spectrophotometer controlled by a personal computer equipped with advanced spectral analysis software was employed. The spectral bandwidth was 2nm and the data acquisition time was 0.5 s. As expected, in both cases the more abundant component (DNA) is accurately measured regardless of the concentration of the less concentrated solute (protein). Importantly, however, multicomponent analysis of the second derivative spectrum yields much more accurate results when the protein absorbance falls below 0.01. The level of detectability was found to be approximately 0.002 absorbance units, which corresponds to approximately 5% w/w for a DNA absorbance of 1.0. These results reflect the widely different absorptivities of protein ($E_{0.1\%, 1cm}$ = 1.0 for an average protein) and DNA ($E_{0.1\%, 1cm}$ = 20). Similar experiments in which DNA was serially diluted into a constant concentration of protein were performed (Fig. 10). Again, the protein concentration was accurately measured at all DNA concentrations. In contrast to dilute protein, however, dilute DNA

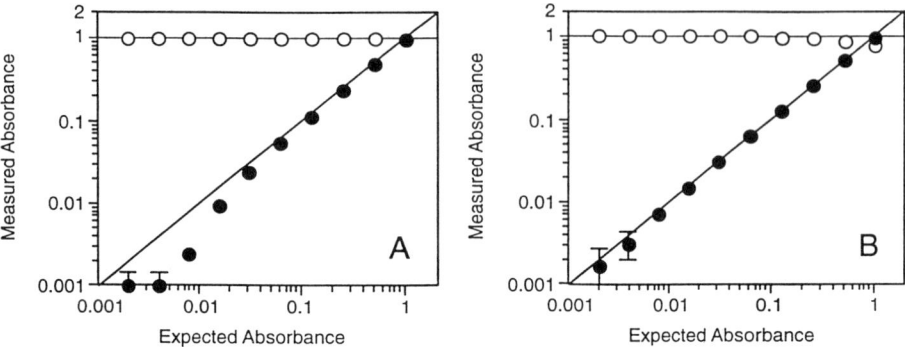

Figure 9.

Multicomponent analysis of protein samples serially diluted into constant concentration of DNA. The protein was detected by monitoring tryptophan absorbance. A: zero-order analysis after light scattering correction. B: second derivative analysis. Open circles denote the 260 nm absorbance values of DNA, while closed circles denote the 280 nm absorbance of protein. Solid lines correspond to the expected positions of data points.

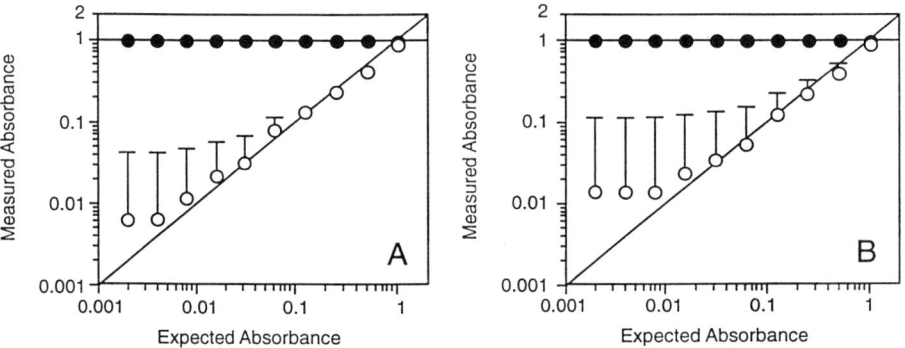

Figure 10.

Multicomponent analysis of DNA samples serially diluted into a constant concentration of protein. A: zero-order analysis after light scattering correction. B: second derivative analysis. Open circles denote the 260 nm absorbance values of DNA, while the closed circles denote the 280 nm absorbance of protein. Solid lines correspond to the expected positions of data points.

was more accurately assayed when a multicomponent analysis of the zero-order spectra (after light scattering correction) was employed. The level of detectability of DNA in this analysis is approx. 0.05 units of absorbance, which correspond to approx. 0.5% w/w. These differences in detection limits no doubt reflect the well defined fine features of the strong second derivative peaks of aromatic amino acids in a protein (Figs. 1-3) compared to the weaker second derivative signals and intense zero-order spectra of nucleic acids (Fig. 4). Traditionally, an A_{260}/A_{280} ratio is employed to assess the purity of DNA samples, with a value of 2.0 assumed for pure DNA (26). We have measured the A_{260}/A_{280} ratio after light scattering correction for 12 globular proteins and found a value of 0.57 +/- 0.06. The equation used to calculate the percent (%) nucleic acid content from this ratio is (26):

$$\%N = (11.16*R - 6.32)/(2.16 - R) \qquad (10)$$

where %N is the weight percent of nucleic acids and R is A_{260}/A_{280} ratio.
It should be noted that application of multicomponent analysis, in addition to yielding an estimate of DNA or RNA content in a protein sample, also provides confidence limits. This safeguards against misinterpretation of increases in the A_{260}/A_{280} ratio due to the appearance of additional non-nucleic acid spectral components.

Second Derivative Spectra as Probes of Protein Structure

The spectral shifts induced by changes in the polarity of the microenvironment of aromatic amino acid chains shown in Figs. 1-3 demonstrate that in well defined native states with fixed neighboring residues, the UV spectrum of a protein should be unique. Structural changes of significant magnitude could potentially result in the perturbation of these native microenvironments and produce corresponding shifts in spectral components. As a consequence of the high reproducibility and sensitivity of diode-array spectrophotometers, such shifts can be quantified to an accuracy and precision approaching 0.01 nm. This is about 1/500 of the total shift expected for the incorporation of the free amino acid into a totally apolar protein core. A convenient way of measuring UV band positions is the calculation of the intersect of the second derivative spectra with the abscissa (X-axis) (Fig. 11). Since the second derivative spectra "oscillate" around the X=0 line, such intersects are independent of protein concentration. Removal of light scattering components or baseline drifts by calculation of second derivatives helps to ensure that these factors do not affect the analysis. Figure 12 shows changes of the value of such intersects during the thermal unfolding of lysozyme. The position of the intersect at 20 °C is a characteristic value for the native state, i.e. changes of this value indicate alteration of the tertiary or secondary structure of a protein or an interaction with a ligand.

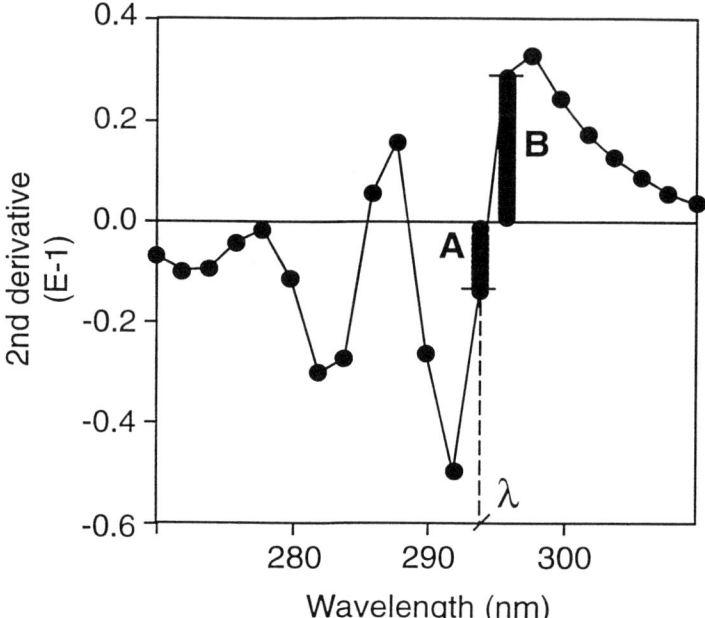

Figure 11.

The second derivative spectrum of lysozyme. The values used to calculate the position of the intersect with the X-axis near 294 nm are shown. A: the absolute value of the second derivative at the data point preceding the intersect, B: the absolute value of the second derivative at the data point following the intersect, L: the wavelength of the data point preceding the intersect. The intersect position is calculated according to the equation: $I(nm) = \lambda + i*A/(A+B)$, where i is the data point interval (nm), and λ is the wavelength of the data point preceding the intersect.

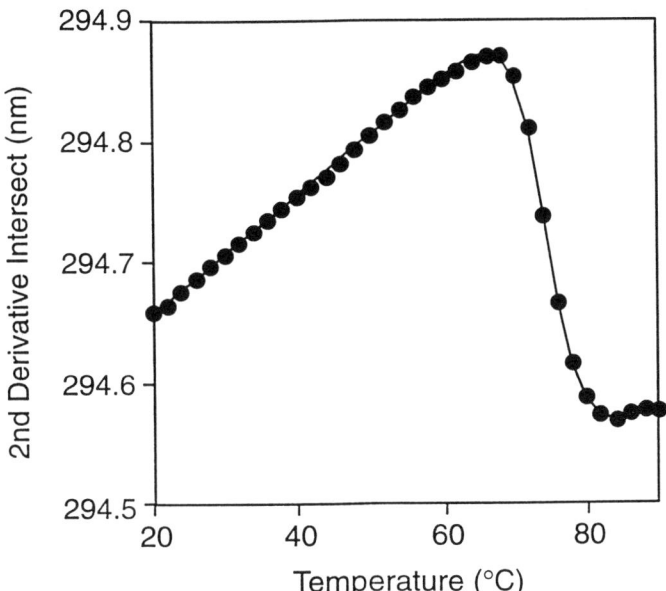

Figure 12.
The position of the intersect of the second derivative spectrum with the abscissa near 294 nm during thermal unfolding of lysozyme. The solid line represents the best fit of the Gibbs-Helmholtz equation after linear fitting of the baselines preceding and following the unfolding transition. Thermal unfolding was induced using a thermoelectric accessory of the HP8452A spectrophotometer. Zero-order spectra (not shown) revealed aggregation occurring at temperatures above 70 °C.

The increasing values of the intersect between 20 °C and 60 °C (red shift) reflect in part the degree of solvent exposure, since the dielectric constant (and thus the polarity) of water decreases with increased temperature. At unfolding temperatures, progressive exposure of the aromatic side chains to the aqueous environment results in a decrease of the intersect position. Fitting of the Gibbs-Helmholtz equation to data in the transition region supports a two-state mechanism of unfolding and yields transition midpoint (T_m) as well as estimates of the enthalpy of unfolding, provided the transition is reversible. It is interesting to note that the precision of these determinations is approximately 0.01 nm, while the data acquisition wavelength interval was 2 nm. This results from the extreme accuracy of the determination of absorbance values at each data point before and after the axis intersect (which are being employed in the interpolation to determine its position). Although a more extensive data base correlating changes of second derivative intersects with changes in far-UV CD and FTIR spectra are lacking, the derivative measurement has the potential to detect tertiary structure changes prior to spectral alterations detected by these more secondary structure-sensitive techniques.

Concluding Remarks

The considerations described above suggest several distinct advantages in applying UV spectroscopy to protein analysis. Diode-array spectrophotometers are inexpensive and reliable, allowing rapid acquisition of highly reproducible spectra. Software is available which allows extraction of valuable quantitative and qualitative information contained in the UV spectra in just a few seconds. The non-destructive nature of these methods also makes them especially convenient for routine protein purity and structural integrity analysis, in addition to concentration determinations. They can be applied to pure samples or to mixtures containing other UV-absorbing components such as nucleic acids, surfactants and stabilizing or antimicrobial agents. In a mixture of DNA and β-lactoglobulin, the level of detectability of protein was found to be approximately 0.002 absorbance units employing second derivative spectra. For the detection of DNA, zero-order spectra yielded a lower limit of detectability (approx. 0.05 absorbance units). Evaluation of the results of multicomponent analysis is aided by parameters describing its reliability, such as the standard deviation of each analyte, relative fit error and independence of standards. The induced unfolding of proteins and nucleic acids can be followed by UV spectroscopy, either to test conformational stability or to screen for effective stabilizing excipients. The position of the intersect of a second derivative spectrum with the wavelength axis was found to be a useful probe of thermal transitions of lysozyme, with spectral resolution approaching 0.01 nm. Information about both tertiary and secondary structure as well as aggregation state of a protein can be revealed in such studies. Diode-array UV detectors are also increasingly used on-line to detect and identify chromatographic peaks, and most of the comments made above are applicable to this situation as well.

Literature Cited

1. Wetlaufer, D.B. *Adv. Protein Chem.* **1962**, 17, 303-390.
2. Ichikawa, T. and Terada, H. *Biochim. Biophys. Acta* **1977**, 494, 267-270.
3. Levine, R.L. and Federici, M.M. *Biochemistry* **1982**, 21, 2600-2606.
4. Servillo, L., Colonna, G., Balestrieri, C., Ragone, R. and Irace, G. *Anal. Biochem.* **1982**, 126, 251-257.
5. Ragone, R., Colonna, G., Balestrieri, C., Servillo, L. and Irace, G. *Biochemistry* **1984**, 23, 1871-1875.
6. Mach, H. and Middaugh. C.R. *Anal. Biochem.* **1994**, 222, 323-331.
7. Mach, H., Volkin, D.B., Burke, C.J. and Middaugh, C.R. (1995) Methods in Molecular Biology, Vol. 40: pp 91-114, *Protein Stability and Folding: Theory and Practice*, B.A. Shirley, ed., Humana Press, Totowa, NJ.
8. Timasheff, S.N. *J. Colloid Interface Sci.* **1966**, 21, 489-497.
9. Volkin, D.B., Tsai, P.K., Dabora, J.M., Gress, J.O., Burke, C.J., Linhardt, R.J. and Middaugh, C.R. (1993) Arch. Biochem. Biophys. **1993**, 300, 30-41.
10. Beaven, G.H. and Holiday, E.R. *Adv. Protein Chem.* **1952**, 7, 319-377.
11. Savitzky, A. and Golay, J.E. *Anal. Chem.* **1964**, 36, 1627-1639.
12. Steiner, R.J., Termonia, Y., and Deltour, J. *Anal. Chem.* **1967**, 44, 1906-1909.
13. Mach, H., Middaugh, C.R. and Denslow, N. (1995) in *Current Protocols in Protein Science*, John Wiley and Sons, New York, Unit 7.2: Determining the Purity and Identity of Recombinant proteins by UV Absorption Spectroscopy.
14. Gratzer, W.B. (1970) in *Handbook of Biochemistry*: Selected Data for Biochemistry (Sober, H.A., ed.), Chemical Rubber Co., Cleveland, pp B 74-77.
15. Mach, H., Middaugh, C.R., and Lewis, R.V. *Anal. Biochem.* **1992**, 200, 74-80.
16. Mach, H., Middaugh, C.R., and Lewis, R.V. *Anal Biochem.* **1992**, 200, 20-26.
17. Cantor, C.R. and Schimmel, P.R. (1980) Absorption Spectroscopy. In *Biophysical Chemistry*, W.H. Freeman, New York, Part II: Techniques for the Study of Biological Structure and Function, pp. 349-408.
18. Lowry, O.H., Rosenbrough, N.J., Farr, A.L. and Randall, R.J. *J. Biol. Chem.* **1951**, 193, 265-275.
19. Bradford, M.M. *Anal. Biochem.* **1976**, 72, 248-254.
20. Perkins, S.J. *Eur. J. Biochem.* **1986**, 157, 169-180.
21. Gill, S.C. and von Hippel, P.H. *Anal. Biochem.* **1989**, 182, 319-326.
22. Pace, C.N., Vajdos, F., Fee, L., Grimsley, G., and Gray, T. *Protein Science* **1995**, 4, 2411-2423.
23. *Hewlett-Packard Instruction Manual*: Understanding Your Advanced ChemStation Software (1995) HP part no G1116-90001, 1st ed., Hewlett-Packard GmbH, Waldbronn, Germany.
24. Mach, H. and Middaugh, C.R. *BioTechniques* **1993**, 15, 240-242.
25. Mach, H., Thomson, J.A., and Middaugh, C.R. *Anal. Biochem.* **1989**, 181, 79-85.
26. Glasel, J.A., *BioTechniques* **1995**, 18, 62-63.

Chapter 12

Surfactant-Stabilized Protein Formulations: A Review of Protein–Surfactant Interactions and Novel Analytical Methodologies

LaToya S. Jones[1,3], Narendra B. Bam[2], and Theodore W. Randolph[1]

[1]Department of Chemical Engineering, Campus Box 424, ECCH 111, University of Colorado, Boulder, CO 80309–0424
[2]SmithKline Beecham Pharmaceuticals, 709 Swedeland Road, King of Prussia, PA 19406

Nonionic surfactants play an important role in the pharmaceutics industry. They are found in purification steps as well final product formulations. Despite the extensive use of nonionic surfactants, their properties, roles and mechanisms by which they yield desired effects are not well understood. This paper discusses the characterization of nonionic surfactants used in pharmaceutics. A review of the binary surfactant - water system provides an introduction to the difficulties encountered when studying more complex systems. Surfactant behavior under formulation conditions, surfactant binding to pharmaceutical products, the role of surfactants in protein refolding, and the effects of surfactants on accelerated testing of formulations is the focus of this review.

Proteins have become increasingly important as pharmacotherapeutics over the past twenty-five years. Consequently, the various commercial processes required for protein production such as cell culture, fermentation, reactor dynamics and protein purification and recovery have been the subject of intense research, and significant headway has been made in the understanding of the basic principles in the fields. However, it should be recognized that a process for production of pharmaceutical proteins must also include storage, shipping, and delivery to the patient. To maintain a biologically active protein during these last three steps, various excipients are commonly added as stabilizers. Understanding the interactions of proteins with these excipients is essential to the rational design of optimal formulations.

[3]Current address: Department of Pharmaceutical Sciences, School of Pharmacy, University of Colorado Health Sciences Center, Denver, CO 80262

Polymers, polyols, and synthetic surface-active agents, surfactants, including nonionic (1-3) and anionic surfactants (4-8) have been traditionally used in formulations to stabilize proteins and, in the case of proteins from blood plasma, to act as antiviral agents (9,10). The utility of surfactants in the biological realm has been formally explored since the late 1930's. It was around this time that Anson, Sreenivasaya, Pirie, and others realized that it was possible to denature proteins using Duponol surfactants (11,12). SDS is a member of this family, and its early presence is perhaps one of the reasons for its dominance in surfactant literature. The behavior and characteristics of this detergent provide insight on the Duponol and other anionic surfactants. However, the surfactants that are used in formulations for their stabilizing properties are normally nonionic. Although these surfactants have been demonstrated empirically to stabilize proteins against long term aggregation and denaturation, the exact mechanism of interaction is often unclear.

In the pharmaceutical industry, a major concern is ease of approval from the regulating body controlling licensing of drug products. An attraction of nonionic surfactants for use in producing, purifying, and stabilizing drugs is that many have already been approved for use internationally in medicinal products. Table I is a list of a few of the approved surfactants. The acceptance is based largely on the general low toxicity and low reactivity with ionic species exhibited by these excipients (13). Surfactants traditionally have been added to protein formulations to increase the protein solubility and/or stability. Recently, pluronic surfactants have been investigated as *in vivo* sustained released vessels for pharmacotherapeutics (14).

Nonionic surfactants are amphipathic molecules containing a bulky polar head group attached to a hydrophobic chain. Chemical sketches of commonly used nonionic surfactants in the pharmaceutical industry are given in Figure 1. It is important to note that many of the structures are characterized by a polydispersity of the hydrophobic chain lengths within a given surfactant. This affects another property of surfactants, the critical micelle concentration (CMC). Surfactant concentrations near the CMC are typically the initial concentrations used in formulation development because it is in this range that properties such as interfacial tension are affected greatly. Also, several of the works referenced in this paper about surfactant - protein interactions discuss trends as a function of surfactant concentrations as related to the CMC.

Determining Critical Micelle Concentration (CMC) and Aggregation Numbers of Pure Micelles

There is a tendency to think of the CMC of a detergent as a constant. However, not only is the CMC dependent on factors such as ionic strength, pH, temperature, surfactant polydispersity, and the presence of other excipients, but the method for determining the CMC can also affect the value. For these reasons, CMCs and corresponding aggregation numbers are best reported as ranges.

Before continuing with methods of determining CMCs and how the above mentioned factors influence the value, a few definitions are necessary. The CMC is defined as the "limiting concentration of single (monomeric) molecules that can exist in solution" (20). Micelles will be defined as organized aggregates of amphiphilic

Table I. Nonionic Surfactants Used in the Pharmaceutical Industry

Chemical Name	Commercial Name	Final Formulation Usage	Quantity	Manufacturer
Polysorbate 20	Tween 20	Actimmune (Interferon gamma-1b)	0.1 mg/ml	Genentech
Polysorbate 40	Tween 40			
Polysorbate 60	Tween 60			
Polysorbate 80	Tween 80	Tubersol (Tuberculin purified protein derivative diagnostic antigen)	0.0005%	Connaught Laboratories
Polysorbate 80	Tween 80	RhoGAM (Rh_0 (D) Immune Globulin)	0.01%	Ortho Diagnostics Systems
Polysorbate 80	Tween 80	Neupogen (Filgrastim)	0.004%	Amgen
Polysorbate 80	Tween 80	Activase (Recombinant Alteplase)	0.11 mg/ml	Genentech
Polysorbate 80	Tween 80	Koate-HP (Factor VIII)	< 25 ppm	Miles Biologicals
Polysorbate 80	Tween 80	Kogenate (Recombinant Antihemopphilic Factor)	<600 µg / 1000 IU	Miles Biologicals
Cetomacrogol 1000	Brij			
Polyethylene Glycol	PEG			

(Adapted from Bam) (15). Final Formulation Usage and Quantity data compiled from *Physicians Desk Reference (PDR)*, 48th Edition, 1994 and is by no means complete. Information regarding specifics of these and other approved excipients for pharmaceutics found in several handbooks (16-19).

$H(OCH_2CH_2)_nOH$

Polyethylene glycol

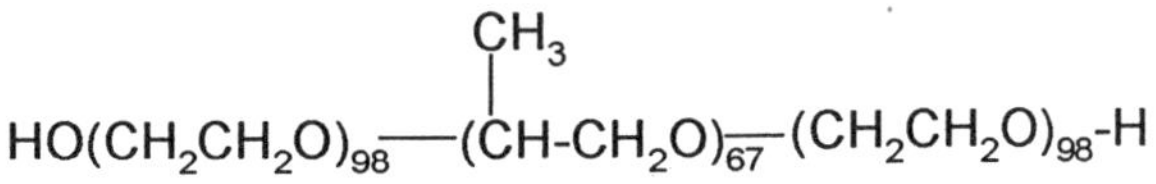

$HO(CH_2CH_2O)_{98}-(\overset{\underset{|}{CH_3}}{CH}-CH_2O)_{67}-(CH_2CH_2O)_{98}-H$

Pluronic F-127

$CH_3(CH_2)_{11}-O(CH_2CH_2O)_{23}-H$

Brij 35

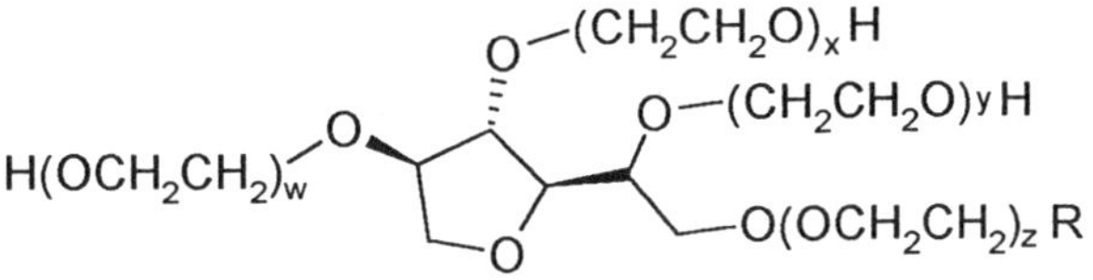

$w + x + y + z = 20$
POLYSORBATE
Tween 20, $R = C_{11}H_{23}CO_2$-
Tween 80, $R = C_{17}H_{33}CO_2$

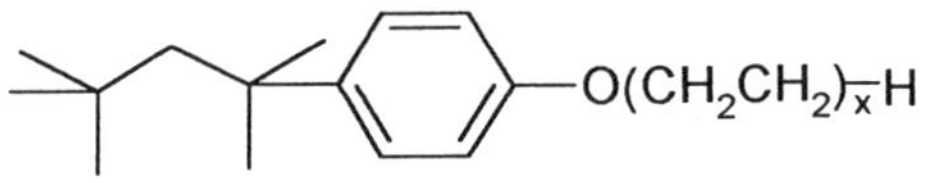

Polyethylene glycol ether
Triton X-100, $x = 9 - 10$ (average)
Triton X-114, $x = 7 - 8$ (average)

Figure 1: Chemical structures of nonionic surfactants of pharmaceutical interest.

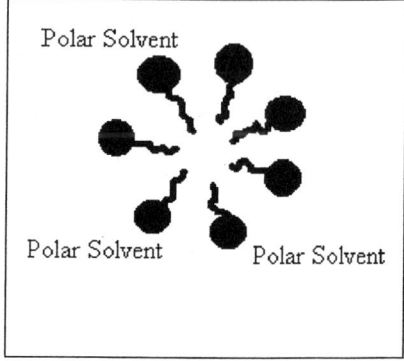

Micelle

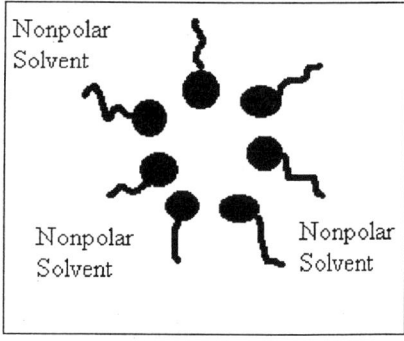

Reverse Micelle

Figure 2: Surfactant Micelle and Reverse Micelle. The hydrophilic group is represented by a circle, whereas the hydrophobic chain is represented by a "squiggle" line.

compounds (21). These "organized aggregates" are oriented such that the hydrophobic portions are in the interior and the hydrophilic parts are exposed to a polar environment. When the hydrophilic portion is in the interior and the hydrophobic portion is exposed to the nonpolar solvent, the structure is termed reverse micelle. Figure 2 is an example of each. Reverse micelles will not be discussed further in this paper; however, Luisi and Laane provide reviews on this topic (22).

Micellization is a positive cooperative process. The micelle formations of concern here are those which form spontaneously, corresponding to a reduction in Gibbs free energy of the system. Micellization reduces the interfacial energy between the water and the hydrophobic tails of the surfactant because these tails are no longer in the aqueous environment (Figure 2). Upon initial consideration, this spontaneous ordering may seem strange. However, for each monomer sequestered as part of a micelle, water molecules which were originally structured about it are now less ordered. The forces that must be overcome for micelle formation include stearic hindrance of bulky head groups, the entropy loss of the surfactant molecule, and for the case of ionic surfactants, charge repulsion. It has been emphasized that the hydrophobic effects observed are the result of the strong attractive forces between water molecules and not due to interactions between hydrophobic moieties of surfactants (21).

Techniques which have been used to determine the CMC of surfactants include: surface tensiometry, refractive index, light scattering intensity, fluorimetry, sedimentation, azo-hydrozone tautomerism, phase selective ac tensammetry, and spin label partitioning (14,15,23-27). In all of these techniques, the CMC is described as the concentration at which there is an initial bend in an isotherm of the measured phenomenon versus concentration. Immediately, it is apparent that the number and range of data points can affect the result. Also, each technique is monitoring a different phenomenon in the solution, with some being more sensitive than others to slight changes. For example, using refractive index and light scattering Kameyama and Takagi determined the CMC for octylglucoside (OG) in distilled water at 22 °C to be 25.3 mM (24). This result was in agreement with the reported value of 25 mM for the CMC of OG at 25 °C determined by surface tensiometry (28). De Grip and Bovee-Geurts used fluorimetry and arrived at a slightly lower CMC of 23.2 mM at 25 °C (29).

It should be noted that OG, when compared to most nonionic surfactants of biological interest, has a relatively high CMC. Under similar conditions for the above OG studies, Tritons and Tweens have CMCs less than 1 mM and 0.1 mM, respectively (30). The consequence of a relatively low CMC combined with a technique in which micellization produces only a slight discontinuity can lead to erroneous results if care is not taken in the experimental design. Too few data points below the CMC coupled with scatter could result in great difficulty in detecting the CMC.

Membrane Solubilization

When therapeutic proteins associated with lipids are formulated with surfactants, the likelihood of lipid membrane solubilization must be considered. Some nonionic

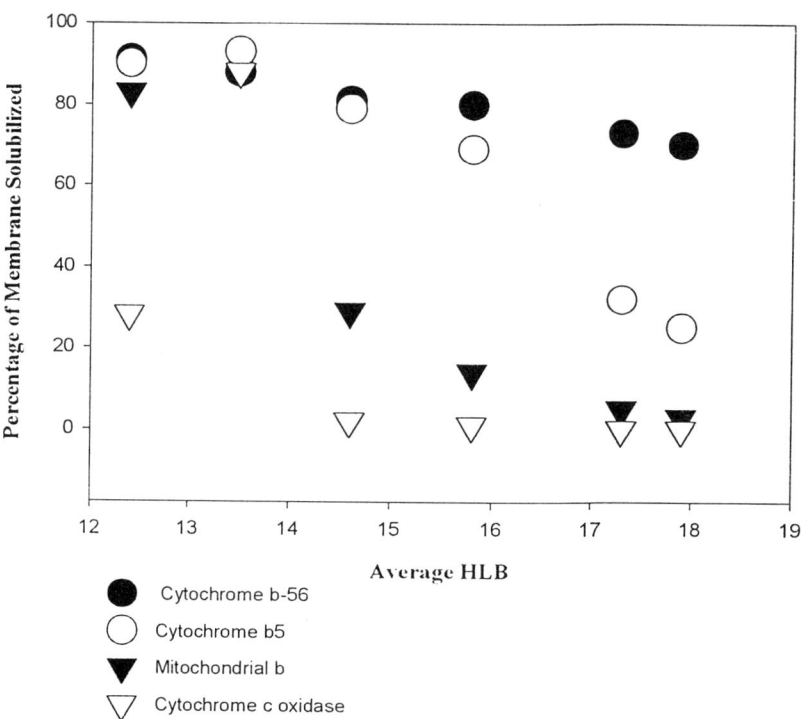

Figure 3: Relationship between HLB and the ability of surfactants from the Triton series to solubilize membranes. (Data from Slinde and Flatmark)(31)

surfactants are quite similar to biological phospholipids. This similarity makes nonionic surfactants good candidates for solubilizing phospholipid membranes. While this membrane solubilization is important in solubilizing integral membrane proteins, it may be undesirable in final formulation steps if lipid association is necessary for proper protein conformation.

Membrane solubilization occurs in the following steps. At low concentrations, some surfactant intersperses in the phospholipid bilayer and the membrane becomes more fluid. As the concentration of surfactant is increased, the surfactant and phospholipids form mixed micelles. The kinetics of solubilization can vary from minutes to hours. The size distribution of these mixed micelles is thought to be Gaussian at equilibrium (21). Studies have shown that when phospholipid - surfactant complexes form, the amount of free monomer in solution is below the pure surfactant CMC (21).

Besides the CMC, Slinde and Flatmark (31) report the consequences of two other properties of surfactants concerning membrane solubilization: the hydrophile - lipophile balance, HLB, and the cloud point. The HLB is a measure of the hydrophilicity of the surfactant, while the cloud point is the solubility limit temperature. To obtain the various HLBs and cloud points for their study, the Triton X series surfactants were used. This minimized the introduction of other variables in the system. They found that membrane solubilization was inversely proportional to the HLB (31) (Figure 3). Empirically, this is what one would expect since hydrophobicity increases with decreasing HLB.

Lichtenberg *et al.* and Jones have provided reviews on the subject of membrane solubilization (20,21). Jones primarily provides thermodynamic information about interactions of SDS with various proteins and biomembranes(20). The review by Lichtenberg *et al.* reviews both surfactant micellization and phospholipid structures as well as the complexes that phospholipids are capable of forming with nonionic surfactants(32).

Surfactant – Protein Binding

In 1968, Pitt-Rivers and Impiombato demonstrated that surfactants are capable of binding to proteins (33). The charged groups of ionic surfactants interact with the charged parts of the protein. A common ionic surfactant is the anionic surfactant, sodium n-dodecyl sulfate (SDS) which is routinely used in SDS - polyacrylamide gel electrophoresis. Here, its ability to bind to sulfhydryl groups is exploited. Although this binding is an interesting example of surfactant - protein interactions, the lack of charges of nonionic surfactants prevents extrapolating these observations to explain the behavior of nonionic surfactant – protein systems.

Binding of nonionic surfactants to proteins and biological macromolecules has been observed (15,34-41). Table II is a list of various techniques which have been used to probe surfactant - protein interactions. We have used the EPR methodology to obtain binding stoichiometries for some serum proteins. Using this spin labeling technique, we were also able to identify specific sites of interaction. Hydrophobic surface sites were required for binding (15,37,38).

Table II. Techniques to Investigate Protein - Surfactant Interactions

Technique	Information obtained
Quantitative equilibrium dialysis	Binding isotherms, Gibbs energy of ligand binding
Molecular sieve chromatography	Binding levels
EPR spectroscopy – spin probes	Binding stoichiometries
EPR spectroscopy – spin labels	Binding stoichiometries, binding sites
Fluorescence	Binding sites
Calorimetry	Enthalpy of surfactant binding and protein unfolding
Ultracentrifugation	Sedimentation coefficients of protein-surfactant complexes, subunit dissociation and molecular weights
Viscometry	Hydrodynamic volume and shape factors, protein unfolding
Static and dynamic light scattering	Molecular weights, diffusion coefficients - complex dimensions
UV difference spectroscopy	Surfactant-induced conformational changes
Neutron scattering	Structure of surfactant - protein complexes
Enzyme kinetics	Surfactant induced enzyme denaturation or activation

Adapted from Jones (20)

Role in Protein Refolding

As more recombinant proteins appear in the pharmaceutical industry, understanding mechanisms of protein refolding becomes increasingly important. The investigation of the role of surfactants in protein refolding stems from earlier research involving the use of molecular chaperones to assist *in vitro*. Recent discoveries have shown that the problems of aggregation and association may be circumvented *in vivo* by "molecular chaperones." These are a class of protein molecules that catalyze correct folding and prevent nonspecific aggregation of newly synthesized proteins in the cell. Specific examples include the chaperonins (42,43), protein disulfide isomerase (44), and peptidyl prolyl isomerase (45). Badcoe *et al.* using lactate dehydrogenase(46) and Martin *et al.* using dihydrofolate reductase (47) have demonstrated a mechanism in which chaperones act by binding to folding intermediates, thus preventing nonspecific aggregation and increasing yields of native active proteins. Protein refolding kinetics are thought to be slowed by misfolding events. Thus, chaperone binding is thought to decrease the occurrence of significant amount of misfolding. However, increasing folding yields in recombinant processes by adding natural chaperonins may not always be practical or even possible. It would involve the co-expression of chaperonins in the cell further complicating recombinant gene expression. Moreover, chaperonins unfold in protein denaturing environments, reducing their practical use under strongly denaturing conditions.

The hypothesis that surfactants could be used to assist protein refolding is based on the similarities between surfactants and phospholipids and the observation that phospholipids help stabilize the structure of integral membrane proteins. As with most studies of protein - surfactant interactions, SDS is again the prevalent detergent in the pioneering work of surfactant assisted protein refolding.

Zardeneta and Horowitz have recently written a review on protein refolding assisted by surfactant and lipids. Included in the review are results from their work with rhodenase, a non-membrane associated protein found in the mitochondria (40). In this case, the above mentioned solubilizing capability of surfactants was exploited to recover precipitated, denatured protein(40,41). An interesting finding was that optimal refolding occurred with surfactant concentrations greater than the CMC but below some empirically determined maximum. Improvements on the yield of native protein were achieved when the micelles which were assisting in the refolding were mixed micelles of surfactant and lipids. Using the mixed micelle system, they were able to recover a native protein where previously the extent of aggregation at high concentration was an impedance (39,40). The intermediate molten globule conformation of the protein is capable of binding to the mixed micelle; however, once the protein is in its native conformation, it can no longer form a complex with the mixed micelle and is released (40).

A related class of chemicals which have been used to assist protein refolding is polyethylene glycol (PEG) (Figure 1) (48-52). This structure forms hydrophilic parts of several nonionic surfactants, and PEG itself exhibits surface activity. Cleland *et al.* have extensively studied PEG - assisted bovine carbonic anhydrase B refolding. The preliminary work in this study revealed that multimers of the first intermediate along the refolding pathway are responsible for slow refolding kinetics and using PEG as a co-solvent is one method of improving the kinetics. However, PEG is capable of

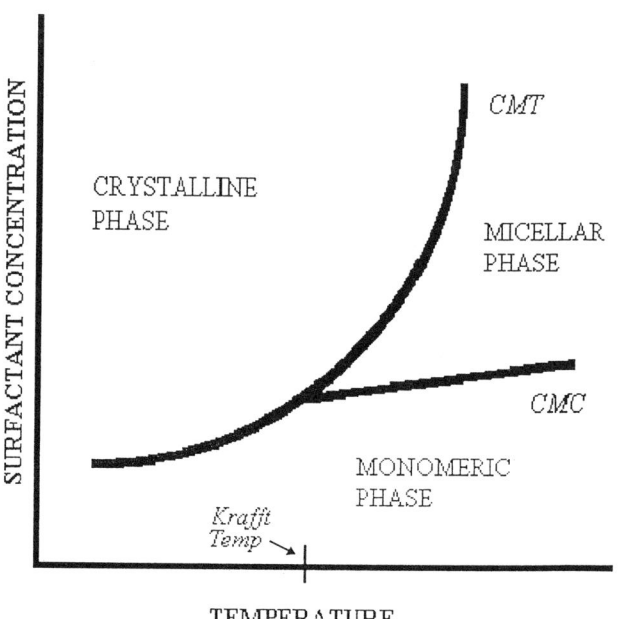

Figure 4: Surfactant Temperature - Composition Phase diagram. (Adapted from Neugebauer) (30)

interacting with proteins by being excluded from the surface due to stearic hindrance or binding to nonpolar regions on the protein(49). Using EPR and fluorescence it was determined that PEG binds to one site of this first intermediate(48). PEG increases the refolding kinetics by preventing aggregation, not by altering the refolding pathway. This model bears similarities to chaperonin studies (50).

Challenges of Accelerated Stability Studies

A common accelerated stability study is one in which protein denaturation is followed as the temperature is raised. Typically, the formulation which stabilizes the protein at the highest temperature is thought to be the best. However, an excipient's inability to protect a protein from thermal denaturation does not necessarily mean that it is incapable of stabilizing the protein. Despite PEG's known stabilizing ability, thermal denaturation studies of proteins in the presence of PEG have yielded results ranging from no effect to a decrease in the T_m of the protein (53). If a formulation decision was based solely on this data, it is likely that PEG would not be added as an excipient. However, the behavior of PEG is dependent on the temperature. PEG tends to bind to the hydrophobic portions of the protein at relatively high temperatures and to be preferentially excluded from the surface at lower temperatures (53).

Temperature strongly affects a surfactant's CMC. Figure 4 is a sketch of a typical temperature - composition phase diagram for surfactants. Notice that in this diagram that instead of the usual solid / liquid / gas phases, the phases of interest are the crystalline / liquid / micelle phases. The Krafft curve gives the solubility limit of surfactant in that it provides a boundary to separate the crystalline phase from the liquid phase (54). The CMT, critical micelle temperature, curve is the boundary for the crystalline and micellar phases. The Krafft point is the triple point defined by the CMC at the CMT (30). As with any phase diagram, the equilibrium composition of each phase for any given temperature and total surfactant concentration can be determined from tie lines (54). If surfactant micelles are responsible for protein stabilization, the temperature - based accelerated stability study can give misleading results. In light of the variation in surfactant CMCs as a function of temperature and the PEG study referenced above, the validity of temperature - induced accelerated denaturation studies must be questioned.

To avoid temperature affects, chemical denaturation can be used instead. Denaturants such as urea and guanidine hydrochloride are used to unfold misfolded and aggregated proteins and in accelerated stability studies. There has been interest in how these additives affect surfactant behavior. For the first use of these denaturants, if protein refolding kinetics are dependent on the presence of micelles, it becomes imperative that micellization trends are known to expedite the optimization of this production step. Also, if surfactants are involved in the stabilizing the final product, any alteration of surfactant properties caused by the addition of chaotropes could decrease the validity of stability studies.

Schick and Gilbert (55) studied the effect of urea, guanidinium chloride, and dioxane on the CMC of branched-chain polyoxyethylene ethers of varying chain lengths. They found that the CMC increased with increasing denaturant concentration. This conforms to the trend observed by Schick for straight-chain polyoxyethylene

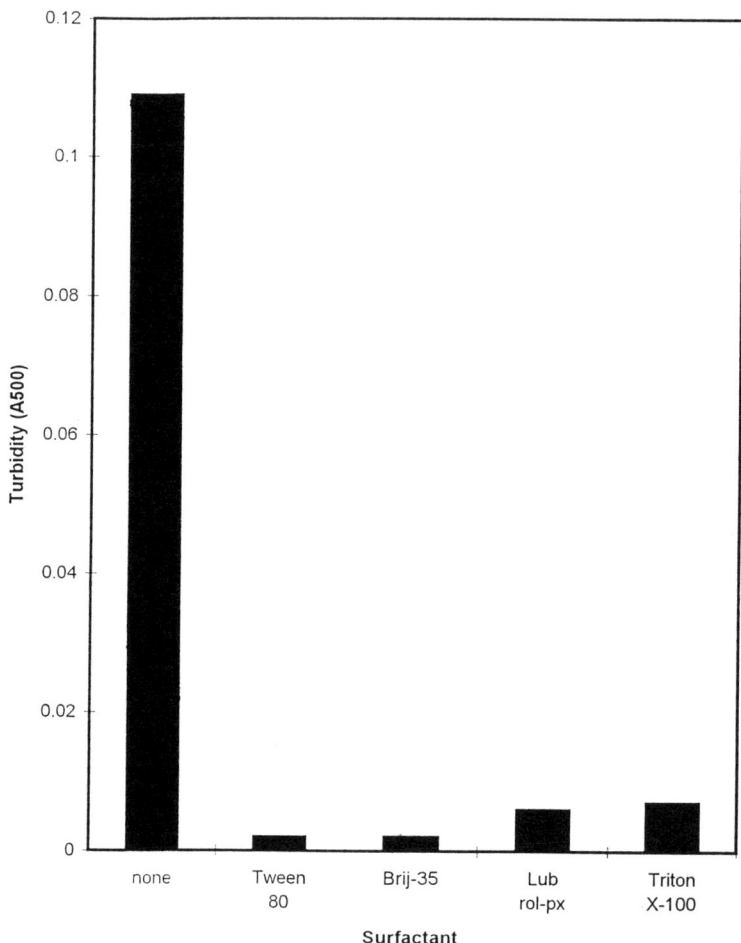

Figure 5: A comparison of the ability of various nonionic surfactants in stabilizing LDH against freeze-thaw induced denaturation. (Data from Chang *et al.*) (61)

ethers and urea (55). Additionally, guanidinium chloride, which is a stronger denaturant than urea, had a greater effect on the CMC (56) (55). Later, Gratzer and Beaven reported that below 3M denaturant the increase in Triton X-100 CMC is greater in the case of urea than guanidinium chloride, but above 3M the opposite trend is observed (57). It should be mentioned that Schick and Gilbert did not investigate chaotrope concentrations between 0M and 3M. Schick *et al.* attributed the increased CMC in the urea and GuCl cases to the disruption of water structure. Other explanations for the effect of dioxane were discussed (55).

From the effect that denaturants and elevated temperatures have on the CMCs of nonionic surfactants, it becomes evident how both chemical and thermal accelerated stability studies can give misleading results if micellization is vital to long-term product stability. Choosing the type of accelerated stability study is not always straightforward and could lead to costly results if the wrong one is selected.

Role in Stabilization against Surface - Induced Denaturation

Proteins are subjected to potential surface - induced denaturation throughout processing and until the point it has finally been administered. With liquid formulations, there is the possibility of shearing stresses caused by agitation and protein coming to the surface and denaturing. In frozen and freeze-dried formulations, the interfaces created by the formation of ice must be considered.

The ability of surfactants to protect blood proteins from interfacial shear stresses has been investigated by Burgess *et al.* (58) The proteins used in this study were bovine serum albumin and human immunoglobulin G. Surface tensiometry was used to measure the kinetics of protein adsorption to three interfaces in the presence and absence of Tween 80 or lecithin. They concluded that these surfactants are capable of preventing protein denaturation at interfaces by winning the kinetic competition of binding at interfaces (58).

Carpenter *et al.* have studied the effects of excipients as cryoprotectants for lactose dehydrogenase (LDH) (59) and phosphofructokinase (PFK) (60). They found that PEG was capable of preventing loses of activity of both LDH and PFK after freeze - thawing. The mechanism of protection was reported to be the Timasheff mechanism of preferential exclusion, which has been used to explain PEG protection in liquid formulation (59).

Chang *et al.* used turbidity as a measure of protein denaturation upon agitation and freeze-thawing (61). In the agitation study, glass beads were used to cause shear stress in liquid formulations. Low concentrations of various nonionic surfactants were capable of protecting the proteins studies from both types of denaturation caused by surface interactions (Figure 5). Although no conclusions were made concerning the mechanism(s) of protection, several possible mechanisms were identified. Among those listed include the above mentioned surfactant-protein binding, surfactant preferentially binding to the interfaces, and surfactant-assisted protein refolding.

Sluzky *et al.* have studied the insulin aggregation denaturation resulting from exposure to hydrophobic surfaces. Insulin denaturation as a function of concentration induced by air - water and Teflon - water interfaces, as well as varying agitation rates, were studied. Although their work did not involve the use of surfactants, it suggested a mechanism for stabilization which could be applicable to systems containing surfactants. Using ultraviolet adsorption spectroscopy, high-performance liquid chromatography, and quasielastic light scattering, they found that the adsorption of the more stable dimeric and hexameric forms of insulin onto hydrophobic surfaces protected the monomeric insulin from interfacial and agitation induced denaturation (62).

Conclusions

The utility of nonionic surfactants in the pharmaceutical industry is evident; however, optimization of their usage is not always straight forward. Surfactants can stabilize protein formulations via several different mechanisms, but it is not always clear which mechanism is at work. Some significant progress has been made in understanding the interactions of proteins and surfactants due to the presence of novel techniques. Through these techniques it has been determined that surfactants are capable of forming complexes with both integral and non-integral membrane proteins. Specific binding sites have even been detected. Also, it is now clear that the consequences of the presence of other excipients as well as the manufacturing conditions must be considered when formulating pharmaceutical proteins.

Acknowledgments

This work was supported by the National Science Foundation (grant # BES 9505301 and a graduate student fellowship) and Merck Corporation.

Literature Cited

1) Chawla, A. S.; Hinsberg, I.; Blais, P.; Johnson, D. *Diabetes* **1985**, *34*, 420-424.
2) Loughheed, W. D.; Albisser, A. M.; Martindale, H. M.; Chow, J. C.; Clemet, J. R. *Diabetes* **1983**, *34*, 424.
3) Twardowski, Z. J.; Nolph, K. D.; McGray, T. J.; Moore, H. L. *American Journal of Hospital Pharmacy* **1983**, *40*, 579-581.
4) Prakash, V.; Nandi, P. K.; Jirgensons, B. *International Journal of Peptide and Protein Research* **1980**, *15*, 305-313.
5) Jirgensons, B. *Journal of Protein Chemistry* **1982**, *1*, 71.
6) Jirgensons, B. *Macromol. Chem. Rapid Commun.* **1981**, *2*, 213-217.
7) Shih, F. F.; Kalmar, A. D. *Journal of Agricultural and Food Chemistry* **1987**, *35*, 672-675.
8) Fukushima, K.; Murata, Y.; Nishikido, N.; Sugihara, G.; Tanaka, M. *Bull. Chem. Soc. Jap.* **1981**, *54*, 3122-3127.
9) Piët, M. P. J.; Chin, S.; Prince, A. M.; Brotman, B.; Cundell, A. M.; Horowitz, B. *Transfusion* **1990**, *30*, 591-598.
10) Snape, T. J. ; Snape, T. J., Ed.; Ellis Horwood: New York, 1991, pp 181.
11) Anson, M. L. *Science* **1939**, *90*, 256-257.

12) Sreenivasaya, M.; Pirie, N. W. *Biochemical Journal* **1938**, *32*, 1707-1710.
13) Lissant, K. J. *Emulsions and Emulsion Technology. part I*; Schick, M. J. and Fowkes, F. M., Ed.; Marcel Dekker: New York, 1974; Vol. 6, pp 440.
14) Kabanov, A. V.; Nazarova, I. R.; Astafieva, I. V.; Batrakova, E. V.; Alakhov, V. Y.; Yaroslavov, A. A.; Kabanov, V. A. *Macromolecules* **1995**, *28*, 2303-2314.
15) Bam, N. B. *Mechanisms of Stabilization of Recombinant Protein Formulations by Surfactants and Polymers*; Yale: New Haven, 1995, pp 226.
16) Smolinske, S. C. *Handbook of Food, Drug, and Cosmetic Excipients*; CRC Press: Boca Raton, 1992.
17) *Handbook of Pharmaceutical Excipients*; American Pharmaceutical Association: Washington D.C., 1986.
18) USPDI *Approved Drug Products and Legal Requirements*; 14th ed. Rockville, MD, 1994; Vol. 3.
19) Reynolds, J. E. F. *Martindale. The Extra Pharmacopoeia*; 29th ed.; The Pharmaceutical Press: London, 1989.
20) Jones, M. N. *Chemical Society Reviews* **1992**, 127-136.
21) Lichtenberg, D.; Robson, R. J.; Dennis, E. A. *Biochimica et Biophysica Acta* **1983**, *737*, 285-304.
22) Zaks, A. *Enzymes in Organic Solvents*; Zaks, A., Ed.; Plenum Publishing Corporation: New York, 1991, pp 161-180.
23) Rosen, M. J.; Cohen, A. W.; Dahanayake, M.; Hua, X.-y. *Journal of Physical Chemistry* **1982**, *86*, 541-545.
24) Kameyama, K.; Takagi, T. *Journal of Colloid and Interface Science* **1989**, *137*, 1-10.
25) Kawashima, N.; Fujimoto, N.; Meguro, K. *Journal of Colloid and Interface Science* **1985**, *103*, 459-465.
26) Müller, E.; Dörfler, H.-D. *Journal of Colloid and Interface Science* **1981**, *83*, 485-494.
27) Palmer, B. J.; Liu, J. *Langmuir* **1996**, *12*, 746-753.
28) Shinoda, K.; Yamaguchi, T.; Hori, R. *Bulletin of the Chemical Society of Japan* **1961**, *34*, 237-241.
29) De Grip, W. J.; Bovee-Geurts, P. H. M. *Chemistry and Physics of Lipids* **1979**, *23*, 321-325.
30) Neugebauer, J. "A Guide to the Property and Uses of Detergents in Biology and Biochemistry," Calbiochem-Novabiochem International, 1994.
31) Slinde, E.; Flatmark, T. *Biochimica et Biophysica Acta* **1976**, *455*, 796-805.
32) Lichtenberg, D. *Biochimica et Biophysica Acta* **1985**, *821*, 470-478.
33) Pitt-Rivers, R.; Impiombato, F. S. A. *Biochemical Journal* **1968**, *109*, 825-830.
34) Patterson, B. W.; Kilgore, L. L.; Chun, P. W.; Fisher, W. R. *Journal of Lipid Research* **1984**, *25*, 763-769.
35) Petri Jr., W. A.; Wagner, R. R. *The Journal of Biological Chemistry* **1979**, *254*, 4313-4316.
36) Van Wijnendaele, F.; Simonet, G. *Method for the Isolation and Purification of Hepatitis B Surface Antigen Using Polysorbate*; Smith Kline-RIT, Belgium: United States, 1987.
37) Bam, N. B.; Randolph, T. W.; Cleland, J. L. *Pharmaceutical Research* **1995**, *12*.

38) Jones, L. S.; Randolph, T. W. *Purification of Recombinant Hepatitis B Vaccine: Effect of Virus / Surfactant Interactions*; Bajpai, R. K., Ed.: University of Missouri-Columbia, 1995, pp 75-82.
39) Zardeneta, G.; Horowitz, P. M. *Analytical Biochemistry* **1994**, *218*, 392-398.
40) Zardeneta, G.; Horowitz, P. M. *Analytical Biochemistry* **1994**, *223*, 1-6.
41) Horowitz, P. M. *Kinetic Control of Protein Folding by Detergent Micelles, Liposomes, and Chaperonins*; Horowitz, P. M., Ed.; American Chemical Society: Washington, DC, 1993; Vol. 526, pp 156-163.
42) Gething, M.-J.; Sambrook, J. *Nature* **1992**, *355*, 33-45.
43) Gatenby, A. A.; Ellis, R. J. *Annual Review of Cell Biology* **1990**, *6*, 125-149.
44) Freedman, R. B. *Protein Disulfide Isomerase: An Enzyme That Catalyzes Protein Folding in the Test Tube and in the Cell*; Freedman, R. B., Ed.; American Association for the Advancement of Science: Washington, D.C., 1991.
45) Lang, K.; Schmid, F. X.; Fischer, G. *Nature* **1987**, *329*, 268.
46) Badcoe, I. G.; Smith, C. J.; Wood, S.; Halsall, D. J.; Holbrook, J. J.; Lund, P.; Clarke, A. R. *Biochemistry* **1991**, *30*, 9195-9200.
47) Martin, J.; Langer, T.; Boteva, R.; Schramel, A.; Horwich, A.; Hartl, F. *Nature* **1991**, *352*, 36.
48) Cleland, J. L.; Randolph, T. W. *The Journal of Biological Chemistry* **1992**, *267*, 3147-3153.
49) Cleland, J. L.; Wang, D. I. C. *Bio/Technology* **1990**, *8*, 1274-1278.
50) Cleland, J. L.; Hedgepeth, C.; Wang, D. I. C. *Journal of Biological Chemistry* **1992**, *267*, 13327-13334.
51) Cleland, J. L.; Wang, D. I. C. *Cosolvent Effects on Refolding and Aggregation*; Cleland, J. L.; Wang, D. I. C., Ed.; American Chemical Society: Washington DC, 1993; Vol. 516, pp 151-166.
52) Cleland, J. L. *Impact of Protein Folding on Biotechnology*; Cleland, J. L., Ed.; American Chemical Society: Washington, DC, 1993; Vol. 526, pp 1-21.
53) Lee, L. L.-Y.; Lee, J. C. *Biochemistry* **1987**, *26*, 7813-7819.
54) Laughlin, R. G. *The Aqueous Phase Behavior of Surfactants*; Academic Press Inc.: San Diego, 1994; Vol. 6.
55) Schick, M. J.; Gilbert, A. H. *Journal of Colloid Science* **1965**, *20*, 464-472.
56) Schick, M. J. *The Journal of Physical Chemistry* **1964**, *68*, 3858-3592.
57) Gratzer, W. B.; Beavan, G. H. *The Journal of Physical Chemistry* **1969**, *75*.
58) Burgess, D. J.; Yoon, J. K.; Sahin, N. O. *Journal of Parenteral Science and Technology* **1992**, *46*, 150-155.
59) Carpenter, J. F.; Crowe, J. H. *Cryobiology* **1988**, *25*, 244-255.
60) Carpenter, J. F.; Prestrelski, S. J.; Arakawa, T. *Archives of Biochemistry and Biophysics* **1993**, *303*, 456-464.
61) Chang, B. S.; Kendrick, B. S.; Carpenter, J. F. *Journal of Pharmaceutical Sciences* **1996**, *85*, 1325-1330.
62) Sluzky, V.; Tamada, J. A.; Klibanov, A. M.; Langer, R. *Applied Biological Sciences* **1991**, *88*, 9377-9381.

Author Index

Bam, Narendra B., 206
Barbieri, David M., 90
Bell, Leonard N., 67
Brocchini, S., 154
Carpenter, John F., 90
Costantino, Henry R., 29
Heller, Martin C., 90
Herman, Alan 168
Hummel, David, 168
Izutsu, Ken-ichi, 109
Jones, LaToya S., 206
Klibanov, Alexander M., 29
Kohn, J., 154
Kojima, Shigeo, 109
Langer, Robert, 29
Liauw, Stanley, 29
Mach, Henryk, 186
Middaugh, C. Russell, 186
Miller, Brian L., 79
Mitragotri, Samir, 29
Prausnitz, Mark R., 124
Randolph, Theodore W., 90, 206
Sanyal, Gautam, 186
Schachter, D. M., 154
Schöneich, Christian, 79
Shahrokh, Zahra, 1
Sluzky, Victoria, 29
Volkin, David B., 186
West, Jennifer L., 119
Wu, Gay-May, 168
Yang, Jian, 79
Yoshioka, Sumie, 109
Zhao, Fang, 79

Affiliation Index

Amgen, Inc., 168
Auburn University, 67
Genentech, Inc., 1
Georgia Institute of Technology, 124
Massachusetts Institute of Technology, 29
Merck Research Laboratories, 186
National Institute of Health Sciences, 109
Rice University, 119
Rutgers, The State University of New Jersey, 154
SmithKline Beecham Pharmaceuticals, 206
University of Colorado, 90, 206
University of Colorado Health Sciences Center, 90
University of Kansas, 79

Subject Index

Activation energies, temperature dependence of protein stability, 3, 4t
Aggregation numbers, determination in pure micelles, 207, 211
Amino acid(s), problems in oxidative degradation, 79–80
Amino acid derived polymers for controlled delivery systems of peptides
available polymers, 155–156, 160f
fabrication of peptide-loaded polyarylate films, 159–161

Amino acid derived polymers for controlled delivery systems of peptides—*Continued*
  peptide release
    compression molded films, 161–163
    polyarylate films, 163–165*f*
  tyrosine-derived polyarylates, 156–159
Arrhenius relationship, temperature dependence of protein stability, 3
Aspartame stability in solids and solutions
  buffer type and concentration, 70–73*f*
  calcium caseinate, 77
  carbohydrate polymers, 77
  carbonyl compounds, 76
  β-cyclodextrin, 77
  degradation kinetics, 68–77
  emulsified oil, 76
  excipient interactions, 76–77
  glass transition, 71, 74*f*, 75
  pathways, 68–69
  pH, 74*f*, 75–76
  temperature, 69–70
  water activity, 71, 73*f*
α-L-Aspartyl-L-phenylalanine-1-methyl ester, *See* Aspartame

Biphasic insulin, development, 36

C-peptide, development, 38
Calcium caseinate, role in aspartame stability in solids and solutions, 77
Carbohydrate polymers, role in aspartame stability in solids and solutions, 77
Carbonyl compounds, role in aspartame stability in solids and solutions, 76
Chemical cross-linking, in situ formation of polymer matrices for localized drug delivery, 122
Chemical degradation pathways, 5–7
Chemical potentials, calculation, 102–106
Compression molded films, peptide release, 161–163
Concanavalin A, use in insulin delivery, 44
Controlled-release drug delivery systems
  amino acid derived polymers of peptides, 154–165
  description, 155

Controlled-release drug delivery systems—*Continued*
  insulin delivery, 42–44
Covalent conjugate characterization, use of laser light scattering photometry, 179–181, 182*f*, 184*t*
Covalent cross-linking, limitations, 119
Critical micelle concentration, 207, 211
β-Cyclodextrin, role in aspartame stability in solids and solutions, 77

Degradable polymer, advantages, 155
Dextran, role in crystallization of components in frozen solutions, 109–117
Diabetes mellitus, 31
Diode-array spectrophotometers, analysis of biopharmaceuticals, 186–204
Drug delivery systems, types, 154

Electroporation, role in modification of skin's barrier properties, 133, 136–137, 139, 143

Fibrillation, stability of insulin, 52–53
Freeze-drying
  physical phenomena, 111
  role in protein stability, 109
Freezing portion of lyophilization, use of polymers for protein stabilization, 90–106
Frozen solutions, phase separation and crystallization of components, 109–117

Glass transition, role in aspartame stability in solids and solutions, 71, 74*f*, 75
Glassy matrix, stability of insulin, 54
Glucose effect
  aspartame stability in solids and solutions, 76
  crystallization of components in frozen solutions, 109–117
Glucose oxidase, use in insulin delivery, 43–44

His–Met peptides, methionine oxidation mechanisms, 86
Human insulin, production, 38

In situ formation of polymer matrices for
localized drug delivery
  chemical cross-linking, 122
  experimental description, 119–120
  photopolymerization, 121–122
  pH-sensitive gelation, 120
  precipitation from organic solutions, 121
  shear-thinning polymers, 121
  thermal gelation, 120
Insulin
  delivery
    aerosol formulations
      nasal delivery, 45–46
      pulmonary delivery, 46
    buccal delivery, 50
    controlled-release devices, 42–44
    ocular delivery, 50–51
    oral administration, 47–49
    parenteral delivery
      continuous subcutaneous insulin
        infusion, 41–42
      implantable intraperitoneal pumps, 42
      jet injectors, 40–41
      pen injectors, 40–41
      subcutaneous injections, 39–40
    rectal delivery, 49
    transdermal delivery, 46–47
    vaginal delivery, 50
  delivery developments, 39–44
  development, 29–30
  examples, 31, 32t
  function, 30
  historical development
    biphasic insulin development, 36
    C-peptide development, 38
    derivative development, 37
    discovery, 33
    early development, 22
    genetic engineering, 38–39
    highly purified insulin development, 36
    pharmaceutical industry development, 5
    production of human insulin, 8
    proinsulin development, 37–38
    regulatory issues, 35
    structure investigations, 36–37
    surfen insulin development, 36

Insulin—*Continued*
  issues in therapy
    delivery, 31
    formulation stability, 33
    patient compliance, 33
  marker, 31
  stability issues, 51–55
  structure, 30
  use in treatment of diabetes mellitus, 31
Insulin-like growth factor, development,
  39
Insulin lispro, development, 39
Iontophoresis
  insulin delivery, 46–47
  role in modification of skin's barrier
    properties, 132–135f

Laser light scattering photometry,
  solution behavior of protein
  pharmaceuticals, *See* Solution behavior
  of protein pharmaceuticals by laser
  light scattering photometry
Liposome effect
  modification of skin's barrier properties,
    127, 129–132
  transport process, 129
Lyophilization, use of polymers for
  protein stabilization, 90–106

Macromolecules, transdermal delivery,
  124–143
Melibiose, role in crystallization of
  components in frozen solutions, 109–117
Met–Met peptides, methionine oxidation
  mechanisms, 86–87
Methionine oxidation mechanisms in
  peptides
  experimental description, 80–81
  His–Met peptides, 86
  Met–Met peptides, 86–87
  neighboring groups, 79–87
  Thr–Met peptides
    hydroxyl radical studies, 81–84
    metal-catalyzed oxidation, 84–86
Micelles, definition, 207, 211
Micellization, description, 210f, 211

Molar absorptivity calculation, UV absorption spectroscopy, 191–192

Neutral protamine lispro, development, 39
Noncovalent aggregation, stability of insulin, 52–53
Nonionic surfactants
 advantages in drug stabilization, 207
 chemical structures, 207, 209f
 examples, 207, 208t
 role in protein stabilization, 206–220
Nucleic acid(s), UV-absorbing chromophores, 187–190
Nucleic acid mixtures, quantitative analysis, 199–201

Oxidation-labile amino acids, problems in oxidative degradation, 79–80

Peptide(s)
 amino acid derived polymers for controlled delivery systems, 154–165
 in situ formation for localized drug delivery, 119–122
 methionine oxidation mechanisms, 79–87
Peptide drugs, clinical effectiveness, 154
pH dependence of protein stability
 aspartame stability in solids and solutions, 74f, 75–76
 chemical-state changes, 5, 6t
 glass container integrity, 5
 physical-state changes, 3, 5
pH-sensitive gelation, in situ formation of polymer matrices for localized drug delivery, 120
Phase separation
 components in frozen solutions, 109–117
 role in formulations, 116–117
Phenylalanine, UV-absorbing chromophores, 188, 190f
Photopolymerization, in situ formation of polymer matrices for localized drug delivery, 121–122
Physical degradation pathways, influencing factors, 7–9

Physical properties of protein solutions, dependence on molecular weight of solute, 168–169
Physicochemical cross-linking, limitations, 119
Pluronics, in situ formation for localized drug delivery, 120
Poloxamers, in situ formation for localized drug delivery, 120
Polyanhydrides, use in controlled delivery systems of peptides, 155
Polyarylate(s), use in controlled-release delivery systems of peptides, 155
Polyarylate films, peptide release, 163–165f
Polycaprolactone, use in controlled delivery systems of peptides, 155
Poly(ethylene glycol)
 in situ formation for localized drug delivery, 120
 phase separation and crystallization, 109–117
 use in controlled-release delivery systems of peptides, 154–165
 use in protein stabilization during freezing portion of lyophilization, 90–106
Poly(glycolic acid), use in controlled delivery systems of peptides, 155
Poly(hydroxybutyrate), use in controlled delivery systems of peptides, 155
Poly(lactic acid), use in controlled delivery systems of peptides, 155
Poly(lactide-co-glycolide), role in stability of insulin, 54–55
Polymer(s) for protein stabilization during freezing portion of lyophilization
 chemical potential relief valve, 95–97, 102–106
 experimental description, 91
 lyophilization process, 91–93
 mechanism of protection, 93–95
 phase splitting during lyophilization, 91
 structural information
  experimental procedure, 98
  Fourier-transform IR spectroscopy, 98
  lyophilization studies, 99–101
Polymer matrices, in situ formation for localized drug delivery, 119–122

Poly(ortho esters), use in controlled delivery systems of peptides, 155
Polyphosphazenes, use in controlled-release delivery systems of peptides, 155
Poly(propylene glycol), in situ formation for localized drug delivery, 120
Polysaccharides, use in controlled-release delivery systems of peptides, 155
Poly(vinylpyrrolidone), role of sugars and polymers in crystallization, 109–117
Precipitation from organic solutions, in situ formation of polymer matrices for localized drug delivery, 121
Proinsulin, 30, 37–38
Protamine zinc insulin, development, 33
Protein(s)
  factors affecting production, 206
  in situ formation for localized drug delivery, 119–122
  subunits, 168
  UV-absorbing chromophores, 187–190
Protein-bound peroxides, production, 80
Protein–ligand interaction characterization, use of laser light scattering photometry, 181, 183–185
Protein mixtures, quantitative analysis, 199–201
Protein pharmaceuticals
  additives for stabilization, 90–91
  factors affecting biological functions, 168
  solution behavior by laser light scattering photometry, 168–185
Protein refolding, role in stabilization, 215, 217
Protein stability, *See* Stability
Pseudopoly(amino acids), use in controlled delivery systems of peptides, 155

Reverse micelle, description, 210f, 211

Shear-thinning polymers, in situ formation, 121
Skin's barrier properties, modification for transdermal delivery of macromolecules, 124–143

Solution behavior of protein pharmaceuticals by laser light scattering photometry
  applications
    covalent conjugate characterization, 179–181, 182f, 184t
    kinetics of molecular interactions, 173, 176–179
    optimal solvent conditions, 173, 175f
    protein–ligand interaction characterization, 181, 183–185
    self-associating systems, 173, 174f
  experimental description, 169
  graphic presentation of experimental results, 171–172
  instrumentation, 170–171
  theory, 169–170
Sonophoresis, insulin delivery, 47
Stability
  analytical tools for assessment, 10–18t
  experimental design considerations, 11
  formulation approaches, 11
  insulin formulations and delivery systems
    chemical stability, 53
    early observations, 51
    physical stability, 52–53
    solid-state stability, 53–55
  pH dependence, 3, 5, 7t
  solids and solutions, *See* Aspartame stability in solids and solutions
  temperature dependence, 2–3
Stabilization, use of polymers during freezing portion of lyophilization, 90–106
Sucrose, role in crystallization of components in frozen solutions, 109–117
Surfactant micelle, description, 207, 210–211
Surfactants, advantages in drug stabilization, 207
Surfactant-stabilized protein formulations
  accelerated stability study challenges, 216f, 217
  critical micelle concentration determination, 207–211
  membrane solubilization, 211–213
  protein refolding, 215, 217

Surfactant-stabilized protein formulations—*Continued*
   stabilization against surface-induced denaturation, 218f, 219–220
   surfactant–protein binding, 213–214
Surfen insulin, development, 36
Synthetic poly(amino acids), use in controlled delivery systems of peptides, 155

Targeted delivery system, description, 154–155
Temperature dependence of protein stability
   activation energies for reaction pathways, 3, 4t
   Arrhenius relationship, 3
   aspartame stability in solids and solutions, 69–70
   conformation stability, 2
   degradation kinetics, 2–3
Temperature-sensitive polymer systems, use in drug delivery, 120
Thermal gelation, in situ formation of polymer matrices for localized drug delivery, 120
Thixotropic polymers, in situ formation for localized drug delivery, 121–122
Thr–Met peptides, methionine oxidation mechanisms, 81–86
Transdermal delivery of macromolecules
   advantages, 124–125
   applications, 125
   experimental description, 124

Transdermal delivery of macromolecules—*Continued*
   modification of skin's barrier properties
   chemical enhancer, 127
   electroporation, 133, 136–137, 139, 143
   iontophoresis as electrical driving force, 132–135f
   liposome, 127, 129–132
   ultrasound, 138–141f, 142t
   pathways across stratum corneum, 125–126, 128f
Tryptophan, UV-absorbing chromophores, 187, 189f
Tyrosine, UV-absorbing chromophores, 188, 189f
Tyrosine-derived polyarylates, use in controlled-release delivery systems of peptides, 156–159

UV absorption spectroscopy for biopharmaceutical analysis
   advantages, 186–187
   light scattering contributions to spectra, 192–195f
   molar absorptivity calculation from protein sequence, 191–192
   multicomponent analysis, 194–199
   quantitative analysis of protein and nucleic acid mixtures, 199–201
   second-derivative spectra as protein structure probes, 201–204
UV-absorbing chromophores in proteins and nucleic acids, 187–190

Valine hydroperoxide, production, 80

Highlights from ACS Books

Desk Reference of Functional Polymers: Syntheses and Applications
Reza Arshady, Editor
832 pages, clothbound, ISBN 0–8412–3469–8

Chemical Engineering for Chemists
Richard G. Griskey
352 pages, clothbound, ISBN 0–8412–2215–0

Controlled Drug Delivery: Challenges and Strategies
Kinam Park, Editor
720 pages, clothbound, ISBN 0–8412–3470–1

Chemistry Today and Tomorrow: The Central, Useful, and Creative Science
Ronald Breslow
144 pages, paperbound, ISBN 0–8412–3460–4

Eilhard Mitscherlich: Prince of Prussian Chemistry
Hans-Werner Schutt
Co-published with the Chemical Heritage Foundation
256 pages, clothbound, ISBN 0–8412–3345–4

Chiral Separations: Applications and Technology
Satinder Ahuja, Editor
368 pages, clothbound, ISBN 0–8412–3407–8

Molecular Diversity and Combinatorial Chemistry: Libraries and Drug Discovery
Irwin M. Chaiken and Kim D. Janda, Editors
336 pages, clothbound, ISBN 0–8412–3450–7

A Lifetime of Synergy with Theory and Experiment
Andrew Streitwieser, Jr.
320 pages, clothbound, ISBN 0–8412–1836–6

Chemical Research Faculties, An International Directory
1,300 pages, clothbound, ISBN 0–8412–3301–2

For further information contact:

American Chemical Society
Customer Service and Sales
1155 Sixteenth Street, NW
Washington, DC 20036

Telephone 800–227–9919
202–776–8100 (outside U.S.)

The ACS Publications Catalog is available on the Internet at
http://pubs.acs.org/books

Bestsellers from ACS Books

The ACS Style Guide: A Manual for Authors and Editors
Edited by Janet S. Dodd
264 pp; clothbound ISBN 0–8412–0917–0; paperback ISBN 0–8412–0943–X

Writing the Laboratory Notebook
By Howard M. Kanare
145 pp; clothbound ISBN 0–8412–0906–5; paperback ISBN 0–8412–0933–2

Career Transitions for Chemists
By Dorothy P. Rodmann, Donald D. Bly, Frederick H. Owens, and Anne-Claire Anderson
240 pp; clothbound ISBN 0–8412–3052–8; paperback ISBN 0–8412–3038–2

Chemical Activities (student and teacher editions)
By Christie L. Borgford and Lee R. Summerlin
330 pp; spiralbound ISBN 0–8412–1417–4; teacher edition, ISBN 0–8412–1416–6

Chemical Demonstrations: A Sourcebook for Teachers, Volumes 1 and 2, Second Edition
Volume 1 by Lee R. Summerlin and James L. Ealy, Jr.
198 pp; spiralbound ISBN 0–8412–1481–6
Volume 2 by Lee R. Summerlin, Christie L. Borgford, and Julie B. Ealy
234 pp; spiralbound ISBN 0–8412–1535–9

From Caveman to Chemist
By Hugh W. Salzberg
300 pp; clothbound ISBN 0–8412–1786–6; paperback ISBN 0–8412–1787–4

The Internet: A Guide for Chemists
Edited by Steven M. Bachrach
360 pp; clothbound ISBN 0–8412–3223–7; paperback ISBN 0–8412–3224–5

Laboratory Waste Management: A Guidebook
ACS Task Force on Laboratory Waste Management
250 pp; clothbound ISBN 0–8412–2735–7; paperback ISBN 0–8412–2849–3

Reagent Chemicals, Eighth Edition
700 pp; clothbound ISBN 0–8412–2502–8

Good Laboratory Practice Standards: Applications for Field and Laboratory Studies
Edited by Willa Y. Garner, Maureen S. Barge, and James P. Ussary
571 pp; clothbound ISBN 0–8412–2192–8

For further information contact:
American Chemical Society
1155 Sixteenth Street, NW ♦ Washington, DC 20036
Telephone 800–227–9919 ♦ 202–776–8100 (outside U.S.)

The ACS Publications Catalog is available on the Internet at
http://pubs.acs.org/books